DOMESTICATION GONE WILD

DOMESTICATION GONE WILD

POLITICS AND PRACTICES OF MULTISPECIES RELATIONS

Heather Anne Swanson, Marianne Elisabeth Lien, and Gro B. Ween, eds.

Duke University Press Durham and London 2018

Printed in the United States of America on acid-free paper ∞
Text design by Courtney Leigh Baker and typeset in Trade Gothic
and Garamond Premier Pro by BW&A Books, Inc., Oxford, North Carolina

Library of Congress Cataloging-in-Publication Data
Names: Swanson, Heather Anne, [date] editor. | Lien, Marianne E., editor. |
Ween, Gro, editor.
Title: Domestication gone wild : politics and practices of multispecies relations /
Heather Anne Swanson, Marianne Lien, Gro B. Ween, eds.
Description: Durham : Duke University Press, 2018. | Includes bibliographical
references and index.
Identifiers: LCCN 2018008900 (print) | LCCN 2018017754 (ebook)
ISBN 9780822371649 (ebook)
ISBN 9780822371335 (hardcover : alk. paper)
ISBN 9780822371267 (pbk. : alk. paper)
Subjects: LCSH: Domestication. | Human-animal relationships. |
Human-plant relationships.
Classification: LCC GT5870 (ebook) | LCC GT5870 .D66 2018 (print) | DDC 392.3—dc 3
LC record available at https://lccn.loc.gov/2018008900

Cover art: Illustration by Per Dybvig

Contents

Acknowledgments

This book began as a workshop called Decentering Domestication, organized by the Department of Social Anthropology at the University of Oslo in November 2014. The event was funded by a research grant from Research Council of Norway on "Anthropos and the Material" and marked the beginning of a sustained interest in domestication at the department. The following year, a ten-month research group called Arctic Domestication in the Era of the Anthropocene at the Center for Advanced Study (CAS), in Oslo, allowed the editors and many of the contributors ample time for revisions and further discussions. We are grateful for the generous and stimulating environment at CAS, which gave us an opportunity to bring together scholars who contributed directly and indirectly to this volume. Special thanks go to David Anderson, Marisol de la Cadena, Diane Gifford-Gonzalez, Britt Kramvig, Kjersti Larsen, John Law, Rob Losey, Andrew Mathews, Gísli Pálsson, Henrik Sinding-Larsen, and Sverker Sörlin for constructive comments and excellent discussions.

In Oslo, various people helped us launch the events that led to this volume. We would like to thank Irene Svarteng, Mette Steensberg, Kristian Sandbekk Nordsted, and Catharina Sletner for keeping track of budgets, people, and practicalities. We thank artist Per Dybvig for providing the drawing on the cover, and CAS, for funding the cost. We also wish to thank our Duke editor, Gisela Fosado, and two anonymous reviewers for their invaluable comments.

INTRODUCTION

Naming the Beast—Exploring the Otherwise ·

Marianne Elisabeth Lien, Heather Anne Swanson, and Gro B. Ween

Concepts of civilization and progress have long been intertwined with the ways people relate to animals and plants, and domestication has been integral to them. Since the nineteenth century, the idea that civilization can be traced to a particular place and time has been central to popular and scholarly imaginaries. As the story goes, civilization emerged from a specific shift in landscape practices, from hunting to husbandry, from gathering to farming. The most studied and discussed period of transition, called the Neolithic Revolution, occurred in the Middle East about ten thousand years ago. This was the transformative moment when human beings allegedly stopped being passively subject to nature and started to be subjects who exerted mastery over it (see, for example, Childe and Clark 1946). Humans cultured themselves by cultivating others, through the domestication of animals and plants. With domestication came a surplus that allowed, but also depended on, larger human settlements to herd the animals, to till the soil, to plant, and to harvest. This, in turn, paved the way for human population growth, division of labor, subjugation of women, social stratification, private property, and state formation. In short, domestication is framed as that which underpins a seemingly inevitable historical road toward "the world as we know it."

This sweeping narrative is compelling and easy to grasp. It is an origin story that explains and orders through binary coupling: the civilized from the savage, the domestic from the wild, progress from regress. Its impact is profound and far from innocent. Closely intertwined with racial and gender hierarchies, colonialism, and the rise of industrial agriculture, this Euro-American story and its many variants (more below) have shaped the worlds we inhabit, as well as our modes of cohabiting with fellow beings. They have sustained and justified biosocial relations that are now hegemonic, such as sedentary agriculture, private property, coercive husbandry, and extractive industries. Positioning Western ways of life as the pinnacle of civilization, domestication narratives have also justified massive interventions from the colonial expansion to the Green Revolution, irreversibly shaping human as well as more-than-human worlds. Hence, the stories told about domestication have served to naturalize and justify a specific and dominant way of life, and they have become political tools in their own right.

As ordering devices, domestication narratives are powerful, because they do not merely classify and divide but also sequence. The categories of "civilized" and "domesticated" are underpinned by stories of domestication that link the Neolithic to the present. The idea that humans began to actively cultivate animals and plants thousands of years ago accentuated and solidified the conceptual separation between culture and nature, offering a tool for ordering people and practices, historically as well as today. The Neolithic Revolution narrative and its various mutations are examples of how classifications are embedded in time: this narrative depicts a watershed historical moment, a linear trajectory of human progress, originating from a particular place and spreading from there to other parts of the world through diffusion or warfare, through conquest or "development." In this way, domestication is integral to the processes through which a Euro-American "natural order of things" emerged.

Why Domestication?

This book explores how situated relations with animals and plants are linked with politics of human difference and, conversely, how politics are historically inscribed in landscapes and seascapes. Seeking to combine insights from multispecies scholarship with critical attention to historically consequential relations of power, we ask: how are the politics of human difference intertwined with plants' and animals' lives, with their changing bodies, and with shifting landscape formations? Rather than writing off domestication as a misinformed narrative or an outdated historical tool, we use it as an entry point into some of the core political stakes and debates that emerge in relation to multispecies anthropology. Through engagement with domestication, we show how plants and animals matter to politics of human difference. In short, we suggest that domestication is a set of ideas ripe for revision at precisely this historical moment.

"Naming the Beast," the first part of this introduction's title, signals that we wish to draw attention to, describe, and ultimately circumscribe domestication narratives in their popular and scholarly forms: their ubiquitous presence in the public realm, their rhetorical effectiveness, and their gaps and silences. Our argument is that even if singular "Neolithic-to-modern" civilization narratives have long been discredited, they continue to haunt, as they speak to foundational concerns about "who we are" as human beings. Asking how popular domestication narratives obscure and shape practices of other-than-human engagement, our first intervention takes the form of critique. Notions of domestication have had far-reaching consequences for colonial and postcolonial politics, nature management, scientific research, and technologies of control and have underpinned an agro-industrial trajectory that is not only socially and politically unjust but also ecologically unsustainable. It is high time to reconsider such structures in light of unexamined assumptions about domestication. This is important not only because domestication narratives have naturalized the dominant environmental practices that "got us into this mess" but also because a critical examination of domestication involves a questioning of narratives we live by. Hence, our intervention is part of an ongoing "decolonialization of thought" (Viveiros de Castro 2011, 128).

The second part of the introduction's title, "Exploring the Otherwise," signals our simultaneous attention to other-than-human practices and relations, both within and beyond the realm conventionally thought of as domestication. Asking how ideas about domestication shape practices of landscape management and human-animal relations, we turn from conceptual critique to ethnographic case studies, and with a specific agenda: seeking out domestication assemblages that are rendered invisible or peripheral by dominant narratives, we explore domestication practices that are thus marginalized, as well as what is commonly seen as the "margins of domestication."

Marginality (like periphery) is constantly made, enacted by narratives as well as practices. Hence, domestication (like modernity) can be seen as a project that constantly produces its own outsides as well as "outsiders within," which, in turn, can be mobilized to justify expanding and civilizing efforts heralded through the idiom of domestication. Through ethnographic attention to domestication assemblages that are marginalized, we can show how multispecies relations become implicated in contexts of colonial expansion, in the making of resource frontiers, and in other efforts associated with progress.

Additionally, thinking domestication through its margins forces us to consider how politics that are justified through idioms of domestication have shaped the margins from which anthropologists conventionally think, such as the nomads, the pastoral, the indigenous, and the remote. Some ethnographic chapters transgress domestication's terrestrial and agricultural biases and seek out practices of domestication that unfold underwater, in air, and in science laboratories. Others challenge the notion of domestication as a spatially bounded practice and draw attention to how culture and the politics of human difference are woven into landscapes and seascapes.

The word "domestication" derives from the Latin *domus*, which in ancient Rome referred to a type of house occupied by the wealthier classes.[1] Dictionary versions link domestication to hearth and home and to the transformation through which something is either converted to domestic use (tamed) or household affairs or made to feel at home (naturalized).[2] Both terms imply the making of insides and outsides through the erection of boundaries, notably between something that is contained within the house, household, or home and something that is not yet contained within that setting.

Yet rather than looking for domestication exclusively "inside the domus" (assuming the "wild," or the "agrios," is elsewhere), this book's contributors draw attention to its blurry boundaries and messy interfaces. Instead of beginning with Middle Eastern grainfields or European pastures, the chapters in this volume take us to unexpected sites of domestication, including Norwegian fjords, Ifugao villages, Japanese forests, falconry cages in Britain, nomadic settlements in Mongolia, and South African colonial townships, where human-animal and human-plant relations exceed the assumptions mobilized by traditional notions of domestication.

Hence, by decentering traditional domestication *narratives*—and recentering *ongoing practices* of domestication—this collection of ethnographic stories contributes to more nuanced understandings of the many kinds of relations that provide continuity and livelihood for human and animal communities. It also shows the great variety of conditions in which how humans relate transform, and are themselves shaped by, their other-than-human surroundings. It also offers alternative ways of imagining our shared futures. Let us turn to the domestication story, as it is conventionally told.

Naming the Beast: The Neolithic Revolution and the Birth of Civilization

While domestication has indeed been narrated in many ways, some versions of the story have proved more charismatic than others. When, in 1928, archeologist Gordon Childe famously coined the term "Neolithic Revolution," he was referring to "that revolution whereby man ceased to be purely parasitic and, with the adoption of agriculture and stock-raising, became a creator emancipated from the whims of his environment" (Childe 1928, 2).[3] In Childe's analysis, revolution was hardly a central concept; in fact, he rarely mentioned it in subsequent writings.[4] But among archeologists and the public, it traveled well, and gradually it became a catchphrase for that Neolithic moment in the Middle East when the history of humanity took a different turn. In this way (and backed by earlier models of unilinear cultural evolution), the Neolithic Revolution came widely to be seen as that watershed moment in which domestication got embedded in time, performing a distinction between the domesticated and the wild, the civilized and the savage.

Childe himself was more concerned with the implication of domestication for the making of Europe. In his books, with popular titles such as *The Dawn of European Civilization* (1925) and *Man Makes Himself* ([1951] 1936), he solidified the idea that progress was a process of enlightenment in which "man" ceased to be a passive prisoner of his environment.[5] Since then, the story has been retold in several ways, and it has been subject to substantial criticism (see Smith 2001; Cassidy and Mullin 2007; Lien 2015). Yet despite such critique, it retains a remarkably strong hold on scholarly and popular imaginations.

While critiquing such narratives is not going to make them go away, critical analysis still remains a useful endeavor in that it draws attention to how they are crafted, to their rhetorical plot, and to their omissions. Such an exercise can strengthen our awareness of the power of popular tropes, making them less smooth, less self-evident, and more open to queries. We have chosen a popular History Channel documentary, *Mankind: The Story of All of Us*, as a convenient example of how domestication often is recounted.[6] We focus on the episode "The Birth of Farming," which is dedicated to the emergence of agriculture and husbandry. The trailer for this episode, readily accessible on YouTube, retells in a remarkable manner the story of the Neolithic Revolution as the route to civilization.[7] It succinctly illustrates how key tropes of classic domestication narratives configure space, time, and agency and how they serve, in turn, to anchor a story of origin for Euro-Americans. Let us turn to the film.

The trailer opens with a pastoral scene: we see a human hand scattering seeds on barren soil and then a glimpse of a long-haired, olive-skinned woman dispersing them while a male voice-over tells us that "*farming is absolutely revolutionary.*" The voice-over then adds, "*When we discovered how to farm, we suddenly increased the ability of the land to support us.*"[8] In the subsequent scene, the woman squats in the middle of a golden grainfield that could be located in the Middle East, a region known to host the first archeological traces of farming, while someone threshes a sheaf of wheat. And yet, evoking a universal "we," the narrator conjures an image of humanity writ large. In this way, the film miraculously presents itself simultaneously as an "origin story" for Euro-Americans and as a "history of mankind." It presents wheat and corn (maize) as the crops that count, conveniently ignoring, for example, Asian histories of rice cultivation or Melanesian cultivation of yams, as well as those Arctic and semiarid regions where the agricultural nexus portrayed in the film was never an option in the first place.

A subsequent scene portrays black men in loincloths wearing face and body paint. They each hold a long spear while running low across a savanna, as if they are stalking game. Accompanied by the sound of African drums, the voice-over states, "*We were very good at hunting.*" But it also reminds us that in some locations, most big animals had died out. With this shift from hunting to farming, the actors' faces shift from black African to white Caucasian, subtly locating the former in a distant past. Through this shift, and by positioning this origin story as the "story of all of us," the film proclaims a notion of domestication that effectively erases those peoples who, by choice or necessity, have relied on other-than-agricultural lifeways, as well as those who have done agriculture differently.

Such erasures help to naturalize and justify Euro-American ideals and present them as universal "goods." "*Our ancestors,*" the voice-over tells us, "*were natural historians.*" We see the woman examining a stalk of grain, touching the seed head with her fingertips. Soon after, we see a close-up of a germinating seed in plowed soil and a time-lapse shot of its rapid development into a seedling rising toward the light. As an increasingly loud symphonic drumroll ushers in this new agricultural world, the voice-over dramatically declares, "*This is the beginning of civilization.*"

The scene then shifts to a monocrop cornfield, husks revealing an ear of corn with perfectly symmetrical deep yellow kernels. The voice-over continues: "*It is the seed from which everything grows. From the first crop to the notion of property. Nation states . . . cities, empires. It is the roots, quite literally of all society."* The accompanying visual images tell a tale of the rise of capitalism and the state: A loaf of leavened bread is changing hands at a rustic market. Coins are being counted and tallied in a ledger book. And then the final image: an animated pastoral landscape, with rolling fields of grain interspersed with patches of green forest, digitally designed so that one literally sees the expansion of cultivated fields like a yellow blanket quickly enveloping most of the terrain.

This final scene brings the point home. Ignoring the often-forced settlement of hunting and gathering peoples by agricultural states (and the formers' resistance), it naturalizes the spatial spread of a singular mode of cultivation: it converts all places into a homogenous landscape of glittering grain, its golden color a sign of its prosperity. The bounty of these fields has ostensibly led to the now-bountiful world of global capitalism and nation-states. The image naturalizes and portrays a world of increasing convergence and an integrated future—a common "modernity" brought into being by farming.

THE HUMAN HAND

The "we" of domestication stories is strongly agential: it is a "we" who discovers, a "we" who builds. Farming, here, is a product not merely of serendipity or interspecies coevolution but of a human intentionality rooted in rational thought. The human hand that scatters the seeds is not an impulsive one but one linked to a thinking being. Agriculture is portrayed as the turning point that changes us from crouched hunters into modern city dwellers. Collapsing thousands of years of multispecies co-evolution into a sequenced tale of necessity, the trailer highlights human strategic action as the key to progress: "*We had to come up with better tools, better houses, to protect our land, to make new materials and so forth. We had new plants to use, we had new plants to grow, we started to develop organized society.*" This emphasis on human intentionality and agency resonates with the idea that through farming, "man makes himself" (cf. Childe's book title 1946). No longer subject to the whims of nature, he has become a subject who acts on a world at his fingertips; he is a man in control.

The video then cuts to a scene of a hut with a simple fence. The people in the scene, now white-skinned with European facial features, perform a diversity of tasks: sorting grain, digging fence postholes, harvesting crop, and overseeing others. The symbol of the human hand is particularly important here. *Homo* hands—with their opposable thumbs—are one of the traits assumed to make humans superior to other animals. We can grasp the world—both to make it and to apprehend it—because we can physically manipulate objects with ease.

With manual agricultural skills come a host of other changes: "*We had to have a hierarchical system that is going to have somebody in charge, to make sure that things are done.*" Humans are positioned in the "driver's seat," capable of controlling their own destiny through actions based on careful cognition. This resonates with common themes in anthropology, in which "man" as the "individualized agent" holds the power to act upon Nature, which is merely acted upon (see, e.g., Descola 2012, 459). Animals, plants, and landscapes are rendered passive, located outside the history of progress, at the mercy of the human hand that conquers, transforms, utilizes, or tames.

PROGRESS TRAJECTORIES

This at once global and Euro-American "we" allows for another conceptual move: the naturalization of "Progress." In the trailer, we see images of a stone wall, a fast, digitized unfurling of what resembles the Great Wall of China, rolling across barren lands while the voice-over states, "*Farming is the most important ingredient in human civilization.*" Domestication stories temporalize as much as they spatialize: farming dominates and expands because it is "more advanced" and "ahead" of other lifeways. Agriculture is a sign of development and improvement—of moving from simple, primitive ways of being to more complex ones, with task specialization, social stratification, and governments capable of large-scale planning.

Naturalizing agriculture as universal progress is underpinned by notions of necessity that, in turn, are used to justify human conflict and warfare. Once the trailer has settled the need to have somebody in charge, another voice continues: "*Farmers are invested in land. Inevitably, if there is more than one person farming, this brings them into conflict with one another.*" Explosively the scene shifts again; the music reverts to African drums, and we see men running, clubs in hand, yelling war cries and aiming at each other with bow and arrow. The voice continues: "*Warfare follows farming as a natural sequence.*" The explanation is simple: "*You have stuff to lose, you have vested interests, and we had to protect other things to protect that.*"

The trailer does not make the shift from protecting one's own land to claiming that of others explicit, but its juxtaposition of images and voice-over, from the animated rollout of a golden blanket of corn to the conjunction of farming and warfare as a natural sequence, evokes a strong message: nonagricultural people are temporally "behind" and immature, while agricultural societies represent the pinnacle of mature human civilization, inevitably expansive, violent, and "naturally" superior. In this way, the story of domestication provides moral justification for colonial and neocolonial projects that "help" so-called underdeveloped peoples, nonagriculturalists, and nomads to attain a higher level of being by teaching them how to approximate Euro-American ideals and agricultural forms.

The trailer described above is a commercial product, made simple and seductive to draw attention to a popular series on the History Channel. Our criticism lies not with the trailer as such. Rather, we seek to draw attention to widespread assumptions that make trailers like this effective. The notion of the Neolithic Revolution is well known. Tropes of such standard domestication stories have shaped both scholarly and popular imaginations across Euro-America and beyond and have served as what Hayden White has called "a practical past"—a popular history that aims to bolster a version of the present and provide guides for futures (2014). The trailer illustrates such tropes, and in this way, it helps us in "Naming the Beast." Our first concern is simply to note its pervasiveness and popular appeal. Our second concern is to consider the work it does toward performing domestication in a particular manner. Through tales like "the birth of farming," we argue that domestication narratives have sustained, justified, and made legible biosocial relations (such as sedentary agriculture, private property, coercive husbandry, and extractive industries) that have had material effects on human and nonhuman lives. These relations have helped to prop up troubling human social formations—including racial hierarchies and the domination of women, patriarchal family structures, reproductive control, naturalized notions of European kinship, and concepts of the household and the domestic—that underpin nation-states as well as imperial colonizing projects. Domestication narratives have also shaped the ways that people have crafted landscapes and forged relations with animals and plants.

Such narratives have also informed scholarly understandings of what domestication is and underpinned approaches to domestication that now seem too narrow. Consider, for example, the much-cited definition of domestication by archeologist Juliet Clutton-Brock, in which domesticated animals are "bred in captivity for purposes of subsistence or profit, in a human community that maintains complete mastery of its breeding, organization of territory and food supply" (Clutton-Brock 1994, 26).[9] "Complete mastery" is a tall order, and as we shall see, it is one that is rarely if ever achieved in interspecies relations, as many studies of people who work closely with animals show (see, for example, Anneberg and Varst, this volume; Lien 2015; Bjørkdahl and Druglitrø 2016; Cassidy and Mullin 2007). And yet, this is the assumption that has been repeatedly evoked in anthropological studies of domestication. Why do we keep forgetting that human lives, bodies, and practices are always deeply entangled other-than-human relations? How could we ignore that culture and politics are lively assemblages in which nonhuman species play key roles? Let us turn to anthropology.

Anthropology and the (Un)making of Domestication Narratives

Anthropologists are well positioned to reconsider concepts of domestication, because their discipline played an instrumental role in hardening classic domestication-as-progress narratives in the first place. Nineteenth-century social scientists, inspired by Scottish Enlightenment thinkers, were the first to explicitly develop the foundations for domestication narratives. Many of these were prominent early anthropologists. Seeking universal laws, Herbert Spencer, Lewis Henry Morgan, and Edward Tylor each worked to formulate versions of what is now called unilinear cultural evolution, a paradigm that included the proposition that cultures move to a higher level of civilization when they adopt settled agriculture. This is Order into History: the sequencing of human difference into a single evolutionary story.[10]

These were not fringe ideas: Edward B. Tylor, for example, is considered one of the founders of social and cultural anthropology and provided the definition of culture that would become the reference point for anthropology as a scholarly discipline (i.e., Boskovic 2004, 524). For Tylor, culture was singular, not plural, and the first sentence of the book *Primitive Culture* states: "Culture, or Civilization, taken in its wide ethnographic sense, is that complex whole which includes knowledge, belief, art, morals, law, custom, and any other capabilities and habits acquired by man as a member of society" (Tylor 1994 [1871], 1). He approached culture as the primary determinant of civilization and held that, through the study of "culture in all its aspects, one could determine the stages that 'mankind' had to pass through in its long quest towards 'civilization'" (Boskovic 2004, 524). Spencer, whose ideas of evolutionary progression from simple to complex society have been equally influential, has been heralded as "the single most famous European intellectual in the closing decades of the nineteenth century" (see, for example, Eriksen and Nielsen 2001, 37). Morgan's work, which included explicit attention to the domestication of plants and grains, both garnered attention in its own right and inspired Marx and Engels, the latter of whom drew on Morgan's texts when he drafted *The Origin of the Family, Private Property, and the State* (1884 [1972]). Morgan's work also inspired Childe (McNairn 1980).

Natural science narratives at once drew on and strengthened such storytelling practices. The biological "tree of life," which typically placed modern (white) man at its crown, embodied a logical structure similar to the era's domestication narratives.[11] In this transdisciplinary moment, ideas of hierarchy, evolution, pedigree, and development traveled between nascent fields such as anthropology and biology, reinforcing these emerging trends and creating a new paradigm of "commonsense."

By the turn of the twentieth century, unilinear cultural evolution had "gone viral" and become a dominant trope, not only among Euro-American intellectuals but also among a variety of publics. Some anthropologists, such as Franz Boas, perceived its dangers and tried to contest its most pernicious parts. He argued against the notion of a singular teleological History and for multiple histories unfolding in different places. Cultural difference, he argued, was spatial, not temporal. People in different locales were equally "developed" to their own unique contexts. No people or way or life was more advanced than any other. While Boas' work helped to temper some of the more pernicious racism in the social sciences, it did not dislodge evolutionary logics. Both within and beyond the academy, stories that focused on developmental sequencing in general—and on the importance of domestication in particular—came to be central to disciplinary imaginations. This centrality was solidified through concepts like the "Neolithic Revolution," which underpins the idea that man is no longer a passive prisoner of his environment's affordances (cf. Childe) but is capable of shaping a world of his own.

Among Boas' concerns was thinking human bodies, their environments, and their social practices *together*. In light of his emphasis on the plasticity of human bodies and the significance of environmental exposures in human development, one might argue that his interventions signaled an early biosocial turn in anthropology (Pálsson 2016, 7). Indeed, canonical anthropological monographs of the twentieth century reflect such concern for corporeal embeddedness and cross boundaries between forms of anthropology that are now held separate, such as physical, cultural, and linguistic forms. Such work posed questions of corporeal coconstitution, such as how human and animal bodies were intertwined through relations of domestication (Evans-Pritchard 1940; Lienhardt 1961; Rappaport 1967).

In the meantime, archeologists considered how domestication might permanently alter bodily features, not only for animals but also for humans (see Leach 2003).[12] Hence, they proposed that physiognomic differences between human groups across the world reflected their adaptation to the environment and, more specifically, whether they were cultivators and/or keepers of domesticated animals.[13] However, the abuse of physical anthropology by the eugenics movement and the atrocities of World War II brought an end to such speculations regarding the connections between human bodies and their relations to animals and plants and has led to what archaeologist Helen Leach has referred to as the "virtual disappearance of this theory of human domestication from post-1950s anthropological writings" (Leach 2003). Along with this disappearance, and as interrelations between landscapes and social and bodily practices were replaced by other concerns (e.g., utility, value, symbol), cross-field conversations among social/cultural anthropology, physical anthropology, and archeology fell silent, too.

By the 1950s, the ontological underpinning of what came to be known as social and cultural anthropology was more or less in place: a notion of culture divorced from physical bodies and evolution. In line with this nonevolutionary program, twentieth-century work on domestication in social and cultural anthropology has highlighted the human side of human-animal relations, focusing on animals as property, on their utility for humans, or on their metaphoric or symbolic meaning (cf. Leach 1964; Lévi-Strauss 1966; see also Cassidy and Mullin 2007).[14] While twentieth-century anthropology produced a number of detailed ethnographic accounts of animal agency, mutualism, and codevelopment within human-animal relations, such descriptions did not pull the materialities of these relations into major theoretical debates in the discipline.[15] Ethnographic descriptions of animal agency and mutualism in human-animal relations, such as that of Evans Pritchard, the Dyson-Hudsons, Lienhardt, Rappaport, or Thompson (see also Fijn, this volume) *could have been* mobilized as a challenge to (then) hegemonic domestication narratives. Instead, they remained secondary to what were held to be these books' main arguments: issues of human social organization and political structure.

We may thus conclude that throughout the latter half of the twentieth century, the Anthropos of anthropology was *not* a figure whose bodily features were shaped by domestication practices, as Boas once suggested (Boas 1911). Rather, it was a figure for whom the body is more or less a given. If her body is malleable, it is through the agency of the human as a thinking and acting subject, rather than the outcome of ancestral practices. The body of social and cultural anthropology is a recipient body that can be acted upon, rather than a dynamic site of interspecies mutuality and evolutionary change (see also Lien 2015, 13). Consequently, until recently, only a handful of scholars engaged in research on the multispecies materialities of domestication. Examples of those who did include Tim Ingold, whose work (see for example Ingold 2000) has challenged standard archaeological and biological definitions of domestication and their emphasis on mastery and control, and David Anderson, who has demonstrated that northern hunters and gatherers, despite their exclusion from agrarian foundational domestication narratives, have maintained complex corelations with reindeer and other animals for more than five thousand years (Anderson 2000).[16] Focusing on Arctic landscapes as cultivated landscapes (rather than remote frontiers of civilization), such contributions draw attention to how such domestication practices involve fine-tuned engagement in the landscape and how human-animal relations imply mutual dependence, dialogue, and trust rather than simplistic forms of control and confinement (Anderson 2000, 2004; see also Willerslev 2008; Losey 2011). Such work shows the promise of what we call "marginal domestications"—the ability of attention to domestication outside of grain-state stories to interrupt classic domestication narratives (see also Lien, this volume). While these authors draw on ethnographic examples from the Arctic, Rebecca Cassidy and Molly Mullin's edited collection *Where the Wild Things Are Now* (2007) widens the scope even more, as they engage the concept of domestication in ethnographic case studies that involve monkeys, lions, farmed salmon, and laboratory mice. These studies have shown how the concept of domestication may yield insight beyond its conventional realms and have demonstrated, above all, that relations of domestication are not always captured by classic notions of human mastery and control. We build on such insights while we take this intervention one step further, through heightened attention to the politics of interspecies relations, as well as the broader implications for the politics of human difference.

In doing so, we draw upon literature that reflects renewed attention to more-than-human relations and to the liveliness of materials, signaling a (new) porosity of disciplinary boundaries. In recent years, literatures on human-nonhuman coconstitution have exploded (see, e.g., Despret 2013; Stépanoff 2012). Terms such as "multispecies ethnography" (Kirksey and Helmreich 2010), "biosociality" (Ingold and Pálsson 2013), "becoming-with" (Haraway 2008), and "more-than-human sociality" (Tsing 2014, 2015) gesture toward a shared commitment to less anthropocentric approaches and framings that permit discussion of human evolution and bodily changes. This volume draws on such insights and explores their political purchase. By foregrounding domestication, we contribute not only to the lively field of more-than-human ethnographic analyses but also to an understanding of how the shifting politics of human difference are profoundly shaped by the legacies of unilinear evolution, domestication narratives, and exclusionary notions of civilization.

The Neolithic Revolution Is Not What It Used to Be

Our approaches to domestication are strongly indebted to insights from archeology, a discipline that has had to deal directly with the troubling afterlives of the primacy of the Neolithic Revolution and its framings of civilizational progress. For several years, many archeologists have been critical of the shortcomings of the domestication narrative, and it is perhaps they who have most consistently challenged the term "domestication." Archeologists Diane Gifford-Gonzalez and Olivier Hanotte (2011) have suggested, for example, that a preoccupation with domestication as an event, or an intention, has produced a lack of curiosity about evolutionary change in domesticated species *after* their first appearance. Such critical reflections resonate with ours and are part of the broader reconsideration of domestication within which our intervention is situated.

With a renewed focus on mutualism and coevolution, the role of nonhuman agency in processes of domestication has been revived (Leach 2003; Gifford-Gonzalez and Hanotte 2011).[17] Equipped with ever-more-sophisticated techniques for reading bones, pots, and pollen (such as genetics and soil analysis), archeologists have scrutinized the role of domestication as a temporal ordering device and have produced new forms of evidence and different imagery for thinking about how domestication evolved. One of the most important insights is that agriculture was not a sudden invention but a long coevolutionary, cumulative process marked by changes in which partner populations of humans and nonhumans became increasingly interdependent (Zeder, Bradley et al. 2006, 139; see also Harlan 1995). Archeologists' main critiques of domestication narratives, which also underpin our own, may be summarized as follows:

1. Domestication is gradual. Acknowledging that the so-called Neolithic Revolution was not a sudden event, some archeologists now prefer the term *neolithisation* to indicate a period in the Near East of at least four to five thousand years when numerous new human-animal-plant regimes occurred (Vigne 2011, 178). Many archeologists also question the dualist temporality of before and after that the very notion of the revolution creates. Hence, they bring attention to what Bruce Smith calls the definitional and developmental "no-man's land" that stretches between hunter-gatherers, on the one hand, and agriculturalists, on the other, a territory that turns out to be "surprisingly large and quite diverse" and that is also difficult to "describe in even the simplest conceptual or developmental term" (Smith 2001, 2).[18]

2 Domestication is not always unidirectional. As this "no-man's land" gets exposed, it turns out not only that the "revolution" took a very long time but also that the historical trajectory of domestication can involve a series of movements back and forth between different food procuring strategies, with shifting emphasis and combination (Gifford-Gonzalez and Hanotte 2011; see also Scott 2011). This is supported by contemporary ethnographic studies of human-animal relations that document how domestication is not a one-way street. Focusing on reindeer among the Evenki, David Anderson details how the collapse of the Soviet state made previously domesticated reindeer "wild," as domesticated herds joined the wild flocks passing by, but also how some Siberian reindeer herding peoples habitually breed in some wild animals (Anderson 2000). Similarly, as Natasha Fijn has argued with reference to the Mongolian horses known as Takhi, wild and domestic should be thought of as fluid states in which there can be considerable crossover, or interbreeding, between the two (Fijn 2011, 2015). Such instances complicate distinctions between wild and domestic and remind us that even though there are many examples of husbandry animals that are genetically altered in ways that make a "return to the wild" unimaginable (a feral poodle could hardly become a wolf), domestication is not always a unidirectional process.[19]
3 Domestication is multiple. Archeological evidence shows that domestication happened independently in several different places and had many different outcomes (Zeder, Bradley et al. 2006; Vigne 2011). Hence, despite its prominence in archeological literature and in popular accounts, the story of the Neolithic Revolution in the Middle East is, in fact, only one of many stories of how people began to cultivate specific crops or raise husbandry animals. Vigne concludes that "very different societies were initiating similar ideas in completely different parts of the world . . . during the same relatively short period of time" (Vigne 2011, 174). One of the sites in which domestication occurred is in Southern Amazonia, where manioc cultivation began six to eight thousand years ago (Zeder, Bradley et al. 2006). Other examples are China (pig domestication ca. 8,000 years ago), Kazakhstan (horse domestication ca. 5,000 years ago; Vigne 2011, 174), and highland South America (llama domestication). Often, these relations are of great cultural significance. Archeological excavations of dog burial sites in Siberia, for example, suggest that dogs were known as distinct persons, requiring mortuary rites similar to those of humans (Losey

2011). Such findings remind us that domestication processes are indeed multifaceted and situated phenomena with many different historical trajectories.

4 Domestication is a mutual process. While studies of domestication have traditionally located humans as the active agent of change, rendering animals, plants, and environmental surroundings as merely acted upon, most scholars today agree that domestication is, at least, a two-way relationship (Russell 2002; Leach 2003; Oma 2010; Zeder 2012).[20] This implies that both or several parties undergo changes (Leach 2003) in which "each species benefits from the other, in terms of its reproductive success" (Gifford-Gonzalez and Hanotte 2011, 4). However, the relation between mutualism, agency, and human intent remains contested: Vigne, for example, insists that human domestications differ from other episodes of mutualisms to a degree that they are no longer coevolution, because humans, via culture, are "able to modify [their] environment according to long-term predictions, fed by a multi-generation memory of successes and failures and using socialized (i.e., flexible) techniques" (Vigne 2011, 177). Others have pointed out that such assumed one-sidedness is problematic, first, because assumptions about human intent may be overstated and the link between intention and outcome is highly uncertain[21] and, second, because animals have fears and desires too, some of which may lead them to seek out human shelter or human-made food (Stépanoff 2012, see also Lien, this volume). Most scholars now agree that dog domestication, for example, was initiated by wolves who began to specialize in feeding within human areas on the remains of leftover meals, particularly hunted animals. These wolves eventually evolved to have less fear of humans, allowing people to more closely interact with them, and human intervention in their breeding did not occur until long after this initial colonization of the human niche by these animals (Coppinger and Coppinger 2001; Zeder 2012). Additionally, in what is called "the social turn" in the life sciences (Meloni 2014) there is a diversity of stories of cultivation practices in nonhuman cultures, such as ants, that go back hundreds of millions of years, indicating that when it comes to gardening, we humans are "late to the game" (Hartigan 2015, 37).

5 Domestication involves transformations that are unintentional and unforeseen. While some archeologists insist that human intent is the distinguishing feature that makes human domestication different

from coevolution between nonhuman species, others take a different view. Archeologist Helen Leach (2007) has argued that although the shifting selective pressures associated with domestication processes have caused major transformations for animals, plants, and people, these selection pressures were *not* under humans' conscious control. Rather, she argues that for most of the time since domestication processes began, "humans have not understood the mechanisms sufficiently to foresee the consequences for the plants and animals that became their focus, let alone appreciate how they themselves have changed" (Leach 2007, 95). At stake here is the relation between human intentions and material effects. There is no doubt that humans have a remarkable capacity for what Vigne calls multigenerational memory of failures and successes and that societies indeed remember (Connerton 1989), making "trial and failure" effective in achieving long-term intended outcomes. Some forms of transformations associated with domestication practices obviously rely on this. But it is also true that many of the effects associated with human domestication practices are unintended. Zoonoses, that is, diseases that spread between animals and humans, were clearly not part of the plan, even if they brought some obvious advantages for colonial empires (Crosby 1986). The contribution of methane gas from cattle in industrial feedlots to climate change is another unintended outcome of current agro-industrial domestication practices, which can hardly be attributed to human intentionality. This effect is significant and a reminder of the lasting and irreversible human impact on the atmosphere and the earth itself, coined as the Anthropocene, a process that, according to some scholars, began with the Neolithic Revolution (Swanson 2016). In other words, while the alleged salience of human regimes of control in relation to domestication practices is not entirely wrong, it is important to keep in mind that such control is only partially achieved in relation to domestication practices in the present and consistently fails to predict the impact of such practices in the future. Put differently, conscious human efforts to exert some kind of control in relation to specific multispecies trajectories should not be conflated with control as an operative mechanism or an outcome in relations of domestication. This calls for a more nuanced approach to control in human-animal relations, one that does not simply treat control as a defining feature that is either absent or present. Above all, these considerations remind us that stories that evoke human mastery over nature as a distinctive feature of domestication are usually too narrow and are often simply wrong.

6 Domestication is a multispecies relation. Although bilateral relations between humans and other species are often highlighted in the domestication literature, recent contributions to the field tell a more complex story. What they teach us, above all, is how domestication of a single species is associated with a host of transformations that involve many other species too, as well as landscapes elsewhere. Environmental historians have shown, for example, how livestock (cattle and sheep) have played a prominent role in early settlers' conquest of North America (Anderson 2004, 152) and Australia (Crosby 1986), partly due to how their presence irreversibly transformed the vegetation and hence the entire colonial landscape. Archeologists have described the rise in infectious diseases associated with the Neolithic Revolution as the first out of three major epidemiological transitions (Barrett, Kuzawa et al. 1998). Alluding to the emergence of zoonoses in the early phases of animal domestication, James Scott has described the microbiological changes associated with contemporary agro-industrial sites as "late neo-lithic multispecies resettlement camps" (Scott 2011, 206), and aquaculture is clearly an example of a similar dynamic (Lien 2017; Law and Lien 2013). These examples remind us that domestication practices reach far beyond the human-animal dyad and challenge us to think differently about their expansive sites (see also Swanson, this volume).

Archeologists have revisited domestication for more than a decade and contributed to much more nuanced understandings of its practices. But even so, the teleological imagery of domestication as a linear trajectory can appear surprisingly stubborn both within and beyond the discipline. From Morgan's ideas of humankind's evolution from savagery through barbarism to civilization, via Childe's captivating image of cultivation as the "dawn of civilization," to contemporary overviews of pathways to domestication among animals and plants, there are clear continuities. Terms like "stages," "pathways," "travelers," "journey," "progress," and "steps" are commonly used (see, for instance, Leach 2007; Zeder 2012; Larson and Fuller 2014). Although many of these terms are clearly metaphorical, they still reinforce an understanding of domestication as a trajectory, and contribute to the naturalization of domestication as destiny and destination. With the image of a journey, it is *as if animals and humans are on their way somewhere* while also engaging in relational practices of biosocial becoming that have no prescribed linearity at all.[22] This is not meant as a critique of archeology as such (archeologists themselves are obviously often aware of these challenges) but rather as a reflection about how concepts are never innocent and how even with the best intentions, progress imagery continues to haunt. Above all, the example reminds us that the idea of domestication as a temporal trajectory is a powerful trope across the disciplines of archeology and anthropology alike.

What happens when one explores domestication beyond such temporal framings? What else do we notice when domestication is decoupled from ideas of progress and growth? If we acknowledge that domestication is a multiple, mutual, and partly unintentional process, how does this change our framing of domestication as an object of study? And what are the implications for an anthropological understanding of the "Anthropos"? These are questions that the contributions to this volume raise and that we explore in more detail below.

Exploring the Otherwise: Domestication as Relational Practice

The domestication narrative we have laid out above helps to clarify the context into which we write, but our project is not primarily one of critique. Rather, we take the concept of domestication, with all its challenges and problematic narratives, as an invitation to cultivate a broad ethnographic curiosity, asking *what else* might be going on under the radar of its popular and triumphant accounts. This is the work of *exploring the otherwise*, to which we gesture in the second half of this introduction's title. What, specifically, does this kind of an ethnographic approach contribute to an understanding of domestication? How does it push us to think more carefully about relations between humans and other-than-humans? Embedded in this volume's ethnographic approach are two important premises: 1) that attention to domestication *practices* must accompany attention to domestication narratives; and 2) that studying domestication from *marginal and atypical sites* may open up different insights than does approaching domestication from its centers. Together, these two ideas make up the approach we call "decentering domestication."

The following three sections of this book decenter domestication in unique ways. Part 1: Intimate Encounters—Domestication from Within, helps us see the unexpected relations that flourish within relatively recognizable domains of domestication. The chapters in this section draw attention to human-animal dyads in settings where human actions certainly shape animal lives—but not necessarily in expected ways. In the encounters of people and other beings within the bars of cages (Schroer), the boundaries of villages (Remme, Fijn), the walls of a pig stall (Anneberg and Varst), or makeshift houses made for wild ducks (Lien), relations are radically different from those portrayed by domestication narratives. We find that even in seemingly enclosed spaces, domestication includes complex boundary work, unexpected intimacies, ontological uncertainties, and bodily coconstitution. By tracing the shifting distributions of agency, this section of the volume asks *what agential capacities* humans and other-than-humans come to have within various domestication assemblages.

Rather than approaching domestication as a crude relation of asymmetrical control, several chapters in part I describe its unexpected combinations of cooperation and coercion and of asymmetry and intimacy. Sara Asu Schroer, for example, shows how courtship rituals of falcons in captivity cautiously unfold between birds and their human breeders. To produce offspring "behind bars," breeders must court falcons on terms that are not their own and that must be constantly negotiated within intense expressions of affection and aggression. Aggression and affection are also present at Danish pig farms, where the pigs themselves come to make unexpected demands on both farmers and animal welfare policies, as described by Inger Anneberg and Mette Vaarst. In this industrial context, human-pig relations are negotiated in material practices, such as the construction of stalls, but this does not preclude an accommodation of agency of the pigs themselves. The distribution of agency among humans and nonhuman animals is uncertain, negotiable, and shifting. Drawing on Bente Sundsvold's rich ethnography about eider ducks on the Northern Norwegian coast (Sundsvold 2010, 2016), Marianne Elisabeth Lien explores how domestication might be seen as a series of generative and tentative interspecies encounters that defy any assumption about human mastery. Together, the ethnographic cases in this section point to the complexity of domestication relations and the varied forms of agency and subjectivity that emerge within them. They demonstrate that domestication practices require a vast amount of coordination to be sustained and that this coordination often depends on the enrollment and active participation of other-than-humans of many kinds.[23]

Part II of this book is titled Beyond the Farm: Domestication as World-Making. Here, we widen our focus both spatially and temporally. Through attention to sites and beings not commonly associated with cultivation and confinement, this section pushes us to ask from *where* one might study domestication.

It is well known that practices of domestication produce rippling ecological effects: cattle rearing changes grasslands, while salmon farming requires the harvest of small fish to produce pelleted feed. However, such processes are generally thought of as external to acts of domestication themselves. The chapters in part 2 place landscape changes at the heart of domestication stories: Here, the "domus"—the scene or site of domestication—is no longer the farm, the pig stall, or the pen. Rather, it is webs of shifting political and ecological relations that weave across the North Pacific Ocean (Swanson), globally connected farms in a remote part of Norway (Hastrup), the air currents of a South African town (Flikke), a series of trout-bearing watersheds (Nustad), and the traveling knowledge practices of a scientific mapping project (Ween and Swanson).

The chapters in this section force us to consider how the *where* of domestication may extend far beyond the farm. Frida Hastrup's chapter, for example, describes how what appears to be an iconic local apple, "naturally" adapted to the steep hillsides of West Norwegian fjord valleys, is a result of geographical wide-ranging connections: grafting experiments of eighteenth century monks, imported agricultural chemicals, and national government subsidies. To stay rooted, the apples require the mobilization of complex networks of support, their domus extending far beyond the locality from which they supposedly originate. Other chapters in this section (see Nustad, Flikke, and Ween and Swanson) move us away from land-based agriculture altogether, asking how water and air are also subject to practices that might productively be explored as forms of domestication. Questions about the sites and scales of domestication are *analytical* as well as empirical interventions, and many of the authors in this volume show how focusing on unusual scenes of domestication leads to new conceptual insights. Rune Flikke, for example, illustrates how the domestication of air—the taming of smells and atmosphere—was central to colonial conquest in Africa. His analysis, which probes how the planting of eucalypts for sanitary purposes was integral to managing racialized fears of disease within the intimacies of colonial encounters, pushes us to imagine the homemaking of settler colonialism in new ways. Rethinking the *where* of domestication also shifts our sense of *who* or *what* is pulled into its relations. The chapters in this section stress that domestication always exceeds dyadic relations between humans and a given species. Definitions of domestication that focus narrowly on the animals and plants that have been intentionally brought under cultivation overlook the many other creatures who find their niches altered, for better or for worse, by domestication practices that may not have targeted them. Heather Anne Swanson's chapter on the wide-ranging effects of Japanese hatchery-bred salmon on the watershed ecologies of Norton Sound, Alaska, is among those that illustrate how domestication can "go wild" and create unexpected transformations across large swaths of land and sea.

The final chapter of this section is a commentary by Anna Lowenhaupt Tsing that explores the ongoing critical potential of domestication. How, it asks, can we continue to talk about problems of confinement, control, and domination at the same time that we acknowledge co-constitution? How might attention to traditional forms of domestication—and their profound violences—be a way to keep inequalities in multispecies conversations? What analytical possibilities open if we limit the definition of domestication to the legacies of the European practices that emerged out of the Neolithic Revolution and see the margins in different terms and as something other than domestication? Tsing's chapter is a reminder to consider the possibilities, as well as the political implications—and limitations—that different approaches to domestication entail.

Politics of Domestication

Tsing's intervention reminds us of the always political nature of conversations around domestication. Domestication practices are *ordering devices* that often rank the civilized and the savage at the same time that they reconstitute temporal cycles and spatial choreographies. Ordering devices are effective political tools that often justify interventions in the name of progress, development, and/or modernity. This is exemplified in Lien's account of how postwar ideals of profitable farming and growth sidelined other, more fragile subsistence relations, including those with eider ducks, undermining the robustness that had historically made life possible. Domestication practices are also implicated in naturalizing spatial/cultural territories such as nations and homelands, through scientific and state practices (as described by Ween and Swanson), as well as through negotiations of wildness and belonging in contested sites (as described by Nustad).

Reengaging domestication is thus also a political act. In anthropology and archeology, domestication is an analytical term that has emerged from the civilizational narratives and landscape-making practices of particular European worlds. Given these histories, how might scholars want to engage this term differently?

This volume does not seek a single answer to this question. Rather, it demonstrates how different ethnographic contexts and concerns call out for different analytical and political projects—and thus different approaches to domestication. Some of the scholars in this volume aim to undermine domestication's power to uphold civilized/savage binaries by expanding the definition of domestication to include human-animal relations that have typically been deemed undomesticated. Lien's work exemplifies this approach and resonates with that of others who work in the Arctic, as it draws attention to unexpected relational intimacies and affordances in landscapes that are otherwise seen as barren or empty (cf. Anderson 2004). If Arctic worlds are shaped by state policies rooted in hegemonic narratives of domestication, then expanding definitions of domestication to make marginal practices more visible is an important mode of challenging those forms of governance.

Other contexts call for other approaches and alert us to how the continued use of a European concept to analyze other worlds might perpetuate the violences of European domination in some situations. Should Ifugao-pig relations necessarily be analyzed through Western categories and terms such as domestication (Remme)? Does the concept of domestication help or hinder our ability to notice the complex relations between people and dogs in places such as Mongolia and Aboriginal Australia (Fijn)? Remme ultimately chooses to expand the concept of domestication to Ifugao practices, while Fijn refuses to do so, instead opting to critique domestication as a European-origin categorization that should not and cannot be expanded to other biosocial worlds.

Tsing, like Fijn, is wary of expanding the concept of domestication; she proposes that anthropologists examine domestication as a particular historical form, rather than use it broadly as a synonym for multispecies relations. Tsing's chapter illustrates how a narrower definition of domestication allows us to better see and critique "the inequalities and intensities of civilization and home" that have been central to projects of patriarchal domination and state control. Several of the chapters in part II of this volume (Flikke, Nustad, and Ween and Swanson) take a similar approach, as they critically examine how colonial and state projects make use of domestication narratives and practices.

The diverse approaches that our volume's chapters display in defining, examining, and writing domestication are central to its overall aim: to illustrate that decentering domestication requires careful consideration not only of human-animal relations but also of the political context and concerns within which those relations are situated. Read as a set, these chapters echo the argument for the book as a whole, as they insist on the importance of reengaging domestication analytically and ethnographically and illustrate the dilemmas, challenges, and strategic choices faced by those who do so. How, in practice, can scholars both "name the beast" and "explore the otherwise"? The following chapters present a range of approaches and partial answers, and we hope that they may also stimulate readers to find their own.

Acknowledgments

Core ideas for this chapter were written during the first author's sabbatical at the University of California, Santa Cruz, in 2012–2013 and presented at the Sawyer seminar at the University of California, Davis, and at Sydney University in spring 2013, and we are grateful for invitations that made this possible. It was subsequently developed during several events within the research group "Anthropos and the Material," at the Department of Social Anthropology at the University of Oslo, and finalized at the Center for Advanced Study (CAS), when the authors were part of the research group Arctic Domestication in the Era of the Anthropocene (2015–2016). The authors are grateful for the generous research funding at CAS that facilitated our collaboration. Special thanks go to Marisol de la Cadena, Natasha Fijn, Rune Flikke, Diane Gifford-Gonzalez, John Law, Rob Losey, Andrew Mathews, Knut G. Nustad, Gísli Pálsson, Vicky Singleton, and Anna Tsing for constructive comments. We also wish to acknowledge the invaluable comments by two anonymous reviewers.

NOTES

1. Ian Hodder (1993) points out how its root is among the most ancient in the Indo-European world. According to Hodder, "domesticate" is linked to the Latin *domus*, Greek *domos*, Sanskrit *damas*, Old Slavonic *domu*, Old Irish *doim*, and the Indo-European *dom- or dem-* (Hodder 1993:45). Hodder further notes how domestication is associated with common English words such as domicile, dominant, dominus, dome, domain, dame, and tame. (See also Tsing, this volume.)

2. "Domesticate: 1. To convert to domestic uses; tame, 2. To accustom to household life or affairs, 3. To cause to be or feel at home; naturalize, 4. To be domestic." Source: Webster's Dictionary (1996).

3. Why "revolution"? For Childe, Karl Marx was a major source of intellectual inspiration, and he "identified with Marxism both emotionally and intellectually" (Trigger 2007). When he wrote this in 1928, we may imagine that the idea of a revolution, for Childe, was a good thing. Hence the adoption of agriculture and husbandry was perhaps, for him, yet another example of an emancipatory moment in which man could rid himself of the shackles of structures that prevented him from reaching his full potential. Childe himself associated this moment with what he calls the "conquest of civilization" (Childe 1928, 2), and it is this idea that seems to be his leitmotif.

4. In fact, even in this citation, he did not put the two words "Neolithic" and "revolution" next to each other. As the word "revolution" hardly ever appears in Childe's own writings (nor in his indexes), it is perhaps a bit unfair that the term "Neolithic Revolution" is so often attributed to him.

5. When Childe wrote this, in the 1920s, man (not woman) was still the appropriate

figure vested with the power to bring progress (for a reappraisal of Childe's contribution, see Tsing, this volume).

6. The documentary series aired in the United States in 2013 and can be viewed online at http://www.history.com/shows/mankind-the-story-of-all-of-us (uploaded September 12, 2016).

7. See https://www.youtube.com/watch?v=bhzQFIZuNFY.

8. Italicized to indicate emphasis in the voice.

9. For a critique, see Cassidy and Mullin (2007).

10. See Foucault (1970).

11. See Ernst Haeckel's "Tree of Life," which he titled "Pedigree of Man" (1879).

12. This idea was also proposed by Darwin, for whom humans, like animals and plants, were subject to natural selection "while at the same time unlike them they themselves practice two other forms of selection, 'unconscious' and 'conscious'" (Leach 2007, 74, citing Darwin 1868 1:214).

13. See Leach (2007), 93–94.

14. Such approaches draw on the work of archeologist Juliet Clutton-Brock, who emphasizes control, captivity, and human profit as key elements (Clutton-Brock 1994).

15. From classic twentieth-century studies like E. E. Evans-Pritchard's *The Nuer* (1940) to Roy Rappaport's *Pigs for the Ancestors* (1967), other-than-humans are frequent ethnographic companions in anthropological accounts. A thorough review of how they have been rendered visible, active, and social (or not) is beyond the scope of this introduction. In this context, it is sufficient to note that with a few important exceptions, there has been a tendency for such interspecies relational practices to be sidelined by other pressing theoretical concerns. Anthropological approaches to African pastoralism, by scholars such as Evans-Pritchard (1940), Rada Dyson-Hudson and Eric A. Smith (1978), and Godfrey Lienhardt (1961), might have been remembered as books about the relational practices involved in the domestication of cows at a particular historical moments, but more often they are referred to for their contributions to anthropological understandings of (human) social organization in stateless societies, segmentary lineage as a structural principle, and cultural adaptions to specific landscapes. Similarly, while Rappaport's account of Melanesian pigs could be read as a story of semidomesticated pigs and their inclusion in ritual activity, it is more often referred to as a book about cultural ecology and the dynamic nature of socioecological equilibrium.

16. Ingold wrote against classic Euro-American definitions of domestication, arguing that they are too closely associated with culturally specific frames of production, technical development, and property while also presupposing ever-increasing human control over the growth and reproduction of plants and animals and ignoring how animals also act upon humans (Ingold 2000).

17. Most archeologists today would argue, for example, that agriculture was not a sudden invention but a long coevolutionary, cumulative process marked by changes on both sides of a relationship in which partner populations became increasingly interdependent (Zeder et al. 2006, 139; Harlan 1995; Smith 1995; Gifford-Gonzalez and Hanotte 2011; see also Cassidy and Mullin 2007). Zoologists Keith Dobney and Greger Larson maintain that the word "domestication" remains too reliant on a strict and mutually exclu-

sive wild-domestic dichotomy and that this has "prevented a deeper appreciation of those animals whose lives are spent somewhere in between" (2006, 269).

18. This has not stopped archeologists from searching for origins. Vigne (2011), who prefers the term "neolithisation" to Neolithic revolution, repeatedly refers to temporal markers like "the birth of domestication" (172), "the birth of animal husbandry" (177), or the Near East as the "cradle of Neolithisation" (177), indicating that a linear evolutionary paradigm and a search for origins still matters.

19. Those who emphasize genetic changes as a sign of domestication might argue that as long as a return to the wild is a possibility, the animal is not fully domesticated in the first place. This approach presupposes that it is possible to distinguish the "wild" from that which is "not wild." A full discussion is beyond the scope of this introduction, but see Lien and Law (2011) for an account of how this plays out in relation to escaped farmed salmon that breed with their distant "cousins" in their ancestors' rivers of origin.

20. For a critique from a nonarcheologist, see Fijn (2011). For counternarratives with a focus on plants, see Scott (2011) and Tsing (2012).

21. Nagasawa et al. (2009) propose, for example, that interspecies bonding between humans and dogs may be enabled by a hormonal feedback loop, mediated by increased levels of oxytocin triggered by eye contact. Neither intentional nor unilineal, such effects can nevertheless significantly shape how the dog-human relation unfolds. Studying reindeer herding systems in Southern Siberia, Charles Stépanoff develops the concept of "reciprocal learning" (Paine 1988) and argues that herders have come to rely on reindeer's cognitive skills, desires, and autonomy to maintain the herd. This does not preclude an asymmetrical relation, but it challenges the distinction between the domestic and the wild. Paradoxically, then, humans can domesticate reindeer only if they keep them wild (Stépanoff 2012, 287, 309).

22. Zeder, for example, is careful to point out that some animals (like cats) "never reached this final destination" (Zeder 2010).

23. See Callon (1986) for more on enrollment.

REFERENCES

Anderson, David. G. 2000. *The Number One Reindeer Brigade: Identity and Ecology in Arctic Siberia.* Oxford: Oxford University Press.

Anderson, David G. 2004. "Reindeer, Caribou and 'Fairy Stories' of State Power." In *Cultivating Arctic Landscapes*, ed. David G. Anderson and Mark Nuttall, 1–17. Oxford: Berg.

Archetti, Eduardo. 1997. *Guinea Pigs: Food, Conflict and Knowledge in Ecuador.* Oxford: Berg.

Barrett, Ronald, Christopher W. Kuzawa, Thomas McDade, and George J. Armelagos. 1998. "Emerging and Re-emerging Infectious Diseases: The Third Epidemiologic Transition." *Annula Review of Anthropology* 27: 247–71.

Bjørkdahl, Kristian, and Tone Druglitrø, eds. 2016. *Animal Housing and Human-Animal Relations.* London: Routledge.

Boas, Franz. 1938. *The Mind of the Primitive Man.* Basingstoke, UK: Macmillan.
Boskovic, Aleksandar. 2004. "Edward Burnett Taylor." In *Biographical Dictionary of Social and Cultural Anthropology*, ed. Vered Amit. London: Routledge.
Callon, Michel. 1986. "Some Elements of a Sociology of Translation: Domestication of the Scallops and the Fishermen of St. Brieuc Bay" In *Power, Action and Belief: A New Sociology of Knowledge*, ed. John Law, 196–223. London: Routledge.
Cassidy, Rebecca, and Molly Mullin, eds. 2007. *Where the Wild Things Are Now: Domestication Reconsidered.* Oxford: Berg.
Childe, V. Gordon. 1928. *The Most Ancient East: The Oriental Prelude to European Prehistory.* London: Kegan Paul, Trench, Trubner.
Childe, V. Gordon. 1951 [1936]. *Man Makes Himself.* New York: New American Library.
Childe, V. Gordon, and Grahame Clark. 1946. *What Happened in History?* New York: Penguin.
Clutton-Brock, Juliet. 1994. "The Unnatural World: Behavioral Aspects of Humans and Animals in the Process of Domestication." In *Animals and Human Society*, ed. A. Manning and J. A. Serpell, 23–36. London, Routledge.
Connerton, Paul. 1989. *How Societies Remember.* Cambridge: Cambridge University Press.
Coppinger, Raymond, and Lorna Coppinger. 2001. *Dogs: A Startling New Understanding in Canine Origin, Behaviour and Evolution.* New York: Scribner.
Crosby, Alfred. 1986. *Ecological Imperialism: The Biological Expansion of Europe, 900–1900.* Cambridge: Cambridge University Press.
Cruikshank, Julie. 2005. *Do Glaciers Listen? Local Knowledge, Colonial Encounters, and Social Imagination.* Vancouver: UBC.
Crutzen, Paul J. 2002. "Geology of Mankind." *Nature* 415, no. 6867: 23.
Crutzen, Paul J., and Eugene F. Stoermer. 2000. "The Anthropocene." IGBO *Global Change Newsletter* 41: 17–18.
Darwin, Charles. 1868. *The Variation of Animals and Plants under Domestication*, vols. 1 and 2. London: John Murray, Albemarle Street. Reprinted by William Clowes and Sons.
Descola, Philippe. 2012. "Beyond Nature and Culture: The Traffic of Souls." *Hau* 2, no. 1: 473–500.
Despret, Vinciane. 2013. "Responding Bodies and Partial Affinities in Human-Animal Worlds." *Theory, Culture, and Society* 30, no. 7/8: 51–76.
Dobney, Keith, and Greger Larson. 2006. "Genetics and Animal Domestication: New Windows on an Elusive Process." *Journal of Zoology* 269: 261–71.
Dyson-Hudson, Rada, and Eric A. Smith. 1978. "Human Territoriality, an Ecological Assessment." *American Anthropology* 80, no. 1: 21–41.
Engels, Friedrich. [1884] 1972. *The Origin of the Family, Private Property, and the State.* E. Untermann, trans. New York: Pathfinder.
Eriksen, G. Thomas, and Finn S. Nielsen. 2001. *A History of Anthropology.* Brighton: Pluto.
Evans-Pritchard, E. Evan. 1940. *The Nuer: A Description of the Modes of Livelihood and Political Institutions of a Nilotic People.* Oxford: Clarendon.

Fijn, Natasha. 2011. *Living with Herds: Human-Animal Coexistence in Mongolia*. Cambridge: Cambridge University Press.

Fijn, Natasha. 2015. "The Domestic and the Wild in the Mongolian Horse and the Takhi." In *Taxonomic Taperstries: The Threads of Evolutionary, Behavioral and Conservation Research*, ed. A. M. Behie and M. F. Oxenham: Australian National University, Canberra, ANU Press.

Foucault, Michel. 1970. *The Order of Things*. New York: Pantheon.

Gifford-Gonzalez, Diane, and Olivier Hanotte. 2011. "Domesticating Animals in Africa: Implications of Genetic and Archaeological Findings." *Journal of World Prehistory* 24, no. 1: 1–23.

Haeckel, Ernst. 1879. *The Evolution of Man*. New York: Appleton.

Haraway, Donna. 2003. *The Companion Species Manifesto: Dogs, People and Significant Otherness*. Chicago: Prickly Paradigm.

Haraway, Donna. 2008. *When Species Meet*. Minneapolis: University of Minnesota Press.

Hard, Jeffrey, Mart R. Gross, Mikko Heino, Ray Hilborn, Robert G. Kope, Richard Law, and John D. Reynolds. 2008. "Evolutionary Consequences of Fishing and Their Implications for Salmon." *Evolutionary Applications* 1, no. 2: 388–408.

Harlan, Jack R. 1995. *The Living Fields: Our Agricultural Heritage*. Cambridge: Cambridge University Press.

Hartigan, John. 2015. *Aesop's Anthropology: A Multispecies Approach*. Minneapolis: University of Minnesota Press.

Helmreich, Stefan. 2009. *Alien Ocean*. Berkeley: University of California Press.

Hodder, Ian. 1993. *The Domestication of Europe: Structure and Contingency in Neolithic Societies*. Oxford: Blackwells.

Ingold, Tim. 2000. *The Perception of Environment: Essays in Livelihood, Dwelling and Skill*. New York and London: Routledge.

Ingold, Tim, and Gisli Pálsson. 2013. *Biosocial Becomings: Integrating Social and Biological Anthropology*. Cambridge: Cambridge University Press.

Kirksey, S. Eben, and Stefan Helmreich. 2010. "The Emergence of Multispecies Ethnography." *Cultural Anthropology* 25, no. 4: 545–76.

Larson, Gregor, and Dorian Q. Fuller. 2014. "The Evolution of Animal Domestication." *Annual Review of Ecology, Evolution, and Systematics* 45: 115–36.

Law, John, and Marianne E. Lien. 2013. "Slippery: Field Notes on Empirical Ontology." *Social Studies of Science* 43, no. 3: 363–78.

Law, John, and Vicky Singleton. 2015. "ANT, Multiplicity and Policy." http://www.heterogeneities.net/publications/LawSingleton2014ANTMultiplicityPolicy.pdf.

Leach, Edmund. 1964. "Anthropological Aspects of Language: Animal Categories and Verbal Abuse." In *New Directions in the Study of Language*, ed. Eric H. Lenneberg, 23–63. Cambridge, MA: MIT Press.

Leach, Helen M. 2003. "Human Domestication Reconsidered." *Current Anthropology* 44, no. 3: 349–68.

Leach, Helen M. 2007. "Selection and the Unforeseen Consequences of Domestica-

tion." In *Where the Wild Things Are Now: Domestication Reconsidered,* ed. R. Cassidy and M. Mullin, 71–101. Oxford: Berg.
Lévi-Strauss, Claude. 1966. *The Savage Mind.* Chicago: Chicago University Press.
Lien, Marianne E. 2015. *Becoming Salmon: Aquaculture and the Domestication of Fish.* Oakland: University of California Press.
Lien, Marianne E. 2017. "Unruly Appetites: Salmon Domestication 'All the Way Down.'" In *Arts of Living on a Damaged Planet,* ed. Anna Lowenhaupt Tsing, Heather Anne Swanson, Elaine Gan, and Nils Bubandt, 107–25. Minneapolis: University of Minnesota Press.
Lien, Marianne E., and John Law. 2011. "Emergent Aliens: On Salmon, Nature and their Enactment." *Ethnos* 76, no. 1: 65–87.
Lienhardt, Godfrey. 1961. *Divinity and Experience: The Religion of the Dinka.* Oxford: Clarendon.
Losey, Rob. 2011. "Canids as Persons: Early Neolithic Dog and Wolf Burials, Cis-Baikal, Siberia." *Journal of Anthropological Archaeology* 30, no. 2: 174–89.
McNairn, Barbara. 1980. *The Method and Theory of V. Gordon Childe.* Edinburgh: Edinburgh University.
Meloni, Maurizio. 2014. "How Biology Became Social, and What It Means for Social Theory." *Sociological Review* 62, no. 3: 593–614.
Mol, Annemarie. 2002. *The Body Multiple.* Durham: Duke University Press.
Morgan, Lewis Henry. 1877. *Ancient Society.* London: Macmillan.
Nagasawa, Miho, Takefumi Kikusui, Tatsushi Onaka, and Mitsuaki Ohta. 2009. "Dog's Gaze at Its Owner Increases Owner's Urinary Oxytocin during Social Interaction." *Hormonal Behaviour* 55, no. 3: 434–41.
Oma, Kristin A. 2010. "Between Trust and Domination: Social Contracts between Humans and Animals." *World Archaeology* 42, no. 2: 175–87.
Paine, Robert. 1988. "Reindeer and Caribou: *Rangifer tarandus* in the Wild and under Pastoralism." *Polar Record* 24, no. 148: 31–42.
Pálsson, Gísli. 2016. "Unstable Bodies: Biosocial Perspectives on Human Variation." In *Biosocial Matters: Rethinking Sociology-Biology Relations in the Twenty-First Century,* ed. Maurizio Meloni, Simon Williams, and Paul Martin, 1–17. Wiley-Blackwell. *Sociological Review* monograph.
Rappaport, Roy A. 1967. *Pigs for the Ancestors: Ritual in the Ecology of a New Guinea People.* New Haven: Yale University Press.
Scott, James. 2011. "Four Domestications: Fire, Plants, Animals and . . . Us." Address delivered at Harvard University, May 4–6, 2011. https://tannerlectures.utah.edu/_documents/a-to-z/s/Scott_11.pdf.
Smith, Bruce D. 1995. *The Emergence of Agriculture.* New York: Scientific American Library.
Smith, Bruce D. 2001. "Low-Level Food Production." *Journal of Archeological Research* 9: 1–43.
Stépanoff, Charles. 2012. "Human-Animal 'Joint Commitment' in a Reindeer Herding System." *Hau: Journal of Ethnographic Theory* 2, no. 2: 287–312.

Strathern, Marilyn. 1992. *After Nature: European Kinship in the Early Twentieth Century*. Cambridge: Cambridge University Press.
Sundsvold, Bente. 2010. "'Stedets herligheter': Amenities of Place; Eider Down Harvesting through Changing Times." *Acta Borealia* 27, no. 1: 91–115.
Sundsvold, Bente. 2016. "Den Nordlandske Fuglepleie": Herligheter, utvær og celeber verdensarv. PhD dissertation. Tromsø: University of Tromsø.
Swanson, Heather Anne. 2016. "Anthropocene as Political Geology: Current Debates over How to Tell Time." *Science as Culture* 25, no. 1: 157–63.
Trigger, Bruce. 2007. *A History of Archaeological Thought*, 2nd ed. New York: Cambridge University Press.
Tsing, Anna Lowenhaupt. 2012. "On Nonscalability: The Living World Is Not Amenable to Precision Nested Scales." *Common Knowledge* 18, no. 3: 505–24.
Tsing, Anna Lowenhaupt. 2014. "More-Than-Human Sociality: A Call for Critical Description." In *Anthropology and Nature*, ed. Kirsten Hastrup, 27–42. New York: Routledge.
Tsing, Anna. 2015. *The Mushroom at the End of the World*. Princeton: Princeton University Press.
Tylor, Edward. 1889. *Anthropology: An Introduction to the Study of Man and Civilization*. London: Macmillan.
Tylor, Edward. 1994 [1871]. *Primitive Culture*. Vol. 1. London: Routledge.
Vigne, Jean- D. 2011. "The Origins of Animal Domestication and Husbandry: A Major Change in the History of Humanity and the Biosphere." *Comptes Rendus Biologies* 334: 171–81.
Viveiros de Castro, Eduardo. 2011. "Zeno and the Art of Anthropology: Of Lies, Beliefs, Paradoxes and Other Truths." *Common Knowledge* 17, no. 1: 128–45.
Willerslev, Rane. 2008. *Soul Hunters: Hunting, Animism and Personhood among Siberian Yukaghirs*. Berkeley: University of California Press.
Zeder, Melinda. 2012. "Pathways to Animal Domestication. " In *Biodiversity in Agriculture: Domestication, Evolution, and Sustainability*, ed. P. Gepts, T. R. Famula, R. L. Bettinger, S. B. Brush, and A. B. Damania, 227–59. Cambridge: Cambridge University Press.
Zeder, Melinda, Daniel G. Bradley, Eve Emshwiller, and Bruce D. Smith. 2006. "Documenting Domestication: The Intersections to Genetics and Archaeology." *Trends in Genetics* 22, no. 3: 139–55.

PART I. INTIMATE ENCOUNTERS

DOMESTICATION FROM WITHIN

1

BREEDING *WITH* BIRDS OF PREY

Intimate Encounters · *Sara Asu Schroer*

The breeding of birds of prey in captivity has long been deemed impossible. In contrast to well-known domesticates such as dogs, horses, and various livestock, birds of prey have resisted human-controlled breeding, a core element of classic definitions of domestication. The territorial and at times aggressive behavior of raptors, portrayed as solitary animals, toward those with whom they are most intimately bonded, namely, their breeding partner or mate, have made them exceptionally difficult to breed under "controlled conditions" within the confines of human-built environments. Yet despite these challenges, since the second half of the twentieth century, the captive breeding of raptors has become a possibility, and various species have been bred successfully through "artificial" insemination as well as through "natural" breeding pairs.[1]

The aim of this chapter is to provide insight into captive breeding practices involving birds of prey, with a focus on the relationships involved in artificial insemination between breeders and "imprinted" breeding birds. I am concerned with how these intimate encounters require rethinking of the question of "control over reproduction," which figures centrally in classic narratives of human-domesticate relationships. Although enclosures and captivity shape breeder-falcon relationships, control is, as I will show, not exercised by humans alone. This is also philosopher of science Vinciane Despret's (2004) point in her investigations of Lorentz's descriptions of engagements with his imprinted jackdaw. Despret argues that rather than focusing on control, we should look for the essence of domestication in "emotional relations made of expectations, faith, belief, trust" (2004, 122). She understands domestication as a practice that has the power to transform both human and nonhuman subjects, opening new identities and ways of relating as a result.

Controlled breeding has been a defining element in classic definitions of domestication for obvious reasons. As this chapter will show, breeding in captivity has indeed dramatically altered the lives of these birds, but not necessarily in ways that completely disempower, alienate, or objectify them. The conditions of breeding birds of prey direct our attention to the choreographies of intimacy and sociality that emerge within such captive breeding practices and that shape not only birds' but also humans' ways of relating to their worlds (Despret 2004; see also Haraway 2008). Because birds are willful creatures with their own desires and rules of conduct, the continuous negotiation of intimacies becomes central for breeders who enter long-term, intimate, and affective social relationships with their birds that must be continuously negotiated.

Captive Breeding and Falconry Practice

Breeding practice cannot be understood without its connection to the practice of falconry. Even though the relationships between humans and birds of prey in falconry practice are diverse and have a long history, these relationships have so far not been considered under the conceptual umbrella of domestication. Like reindeer-pastoralist relationships in the Circumpolar North, sometimes referred to as "weak" forms of domestication or "semi-domestication" (Ingold 1994), the dynamic relationships involved in falconry do not easily fit into stark categories of the wild and the tame, nor do they lend themselves to domestication narratives of human superiority, domination, and control. In falconry, humans and birds of prey learn to cooperate with each other in the task of hunting. The practice involves methods and techniques peculiar to falconry and involves processes of taming and training, which in previous research I have analyzed in terms of colearning and enskilment (Schroer 2015, 2018).

Traditionally, falconers derived their birds from the "wild," where they were either taken as nestlings from the eyrie (eyasses), trapped on their first migration (passage hawks), or taken as adult birds (haggard). In this context, the "wildness" of birds of prey is not understood as something to be contested or overcome but rather something falconers appreciate and strive to get close to. In contrast to other human-animal relationships in which captive breeding led to the development of animal breeds that in some cases are considered superior to their "wild" relatives, in falconry this longing for improvement of bird species through human interference did not become dominant (for a contrast, see, for example, Cassidy 2002 in relation to the breeding of racehorses). Historically, when the taking of wild birds was still allowed, falconers typically did not put much effort into the breeding of birds in captivity, which has often been considered a difficult and time-intensive enterprise.

Today, the taking of wild birds is illegal or highly regulated in most countries, and the captive breeding of birds of prey has become a regular way for falconers to continue their practice. The first large-scale breeding successes with raptor species began to emerge in the 1960s. During this time, people concerned about birds of prey were alarmed by the decline of some species of raptors—some to the brink of extinction—due to the use of pesticides in industrial agriculture (Ratcliffe 1980). This resulted in joint efforts by falconers and researchers in different countries to work on the captive propagation of birds of prey to produce offspring for reintroduction into the wild. Much of this restoration work has been based on knowledge from falconry practice, using techniques of handling, taming, training, and hunting with birds of prey to create adequate domestic breeding environments and to keep the birds in healthy condition (both physically and mentally). During this time, much was achieved, not only by research institutions but also by private falconer-breeders who experimented with breeding birds of prey in captivity. Breeding success was based on several improvements in relation to the birds' diets and built environment and on a better understanding of the birds' complex courting rituals. It furthermore benefitted from experimenting with artificial insemination and the ability to sex birds more accurately through DNA testing (Weaver and Cade 1991; Platt et al. 2007; Roots 2007).

For many falconers in Britain who have the time and financial resources to do so, the breeding of their own birds has become an extension of their falconry practice. Some falconers welcome it as a way to learn more about their birds. Breeding provides the opportunity for them to be included in the processes of pair bonding and courtship as well as the early stages of a young bird's development. This is usually not the case when they purchase already-flying youngsters from commercial breeders. Because the creation of a social bond between human and bird is central for developing hunting cooperation later on, it becomes important to know that young birds did not have bad experiences under the breeder's care that might lead to a negative association with human beings. Falconer-breeders routinely emphasized that working with birds of prey requires being responsive to interspecies forms of communication and rules of conduct. The falcons, hawks, or eagles with whom one hunts or breeds are usually understood as distinctly willful creatures that can show affection and cooperation toward humans as much as they can dominance and aggression. Falconers, in turn, do not understand themselves as dominant masters of inferior beings but experience their engagements with the birds as cooperative. Attempts to use force will likely make a bird unwilling to cooperate, thereby making its continued domestication an impossible project. This is a point that I will further explore in the next section.

"Controlled" Breeding of Birds of Prey

At first glance, the practices involved in the captive breeding of birds of prey based on selective breeding and the "production" of offspring for either private, conservational, or commercial purposes seem to be at the heart of some of the classic narratives of domestication that prevail in predominantly Western discourses on the human-domesticate relationship. Indeed, in many such discourses, controlled domestic breeding—the removal of animals from the spaces to which they "naturally" belong and their integration and subjugation into the human domain or household—is depicted as "the essence of domestication" (Bökönyi 1989 22; 1969). In these narratives of cultural progress, the human is seen as standing in a dominant and superior position to the domesticated animal. For example, Clutton-Brock argued that, for domestic creatures, the human community "maintains *total* control over their breeding, organization of territory, and food supply" (Clutton-Brock 2012, 3, emphasis added; see also introduction).

A focus on the dynamic interspecies interface within the captive breeding of birds of prey helps to challenge and decenter such classic understandings of domestication that appear too generalizing to capture the manifold relationships and practices that constitute avian domestication. First, drawing a strict boundary between the lives of the captive animals, on the one hand, and the "human community," on the other, seems difficult to achieve in the case of falcon breeding. The practice involves continuous social contact between humans and birds who gradually learn how to engage in meaningful communication in which both take part in creating "rules" for their social interactions. This relationship, then, challenges the human-animal divide that attributes subjectivity to humans only and encourages thinking about the significance of sociality in the domestication process. The "social" here is not a domain bound to a species; rather, in captive breeding, sociality should be understood as an emergent process through which living beings grow in relation to each other and their environments (Ingold 2000, 2011; see also Long and Moore 2012; Strathern and Toren, in Ingold 1995, 50–55, 60–63; and Tsing 2013 on "more-than-human sociality").

Second, the notions of control and dominance must be understood in their various enactments, rather than as purely human and unidirectional (see Anderson et al. 2017; Sax 2014 on Porcher). Captivity is, of course, what allows for intimate human-bird relations, and in the end, falcon breeding is, in some ways, unequal and even possibly coercive. Yet, while breeders acknowledge that they put falcons in aviaries, they also stress that the birds' actions shape and contribute to human-animal engagements within these captive environments. When one pays attention to the experiences and narratives of breeders, it becomes evident that birds are perceived by the breeders as taking active part in, and influencing, breeding practice and their engagements with humans. This is particularly the case when it comes to the protection of their territories that are formed through the birds' enactments of dominance toward potential mates (e.g., a bird or the breeder) as well as toward outsiders that may try to intrude (e.g., an anthropologist on fieldwork trying to clean their aviary, whom they stoop-attacked from above). Furthermore, while breeders may have control over the amount and quality of food they supply, how this food is accepted and consumed by the birds lies outside their control. Especially during the time of courtship, food and mutual feeding have an important meaning for the establishment of a pair bond between breeder and imprinted birds. Before considering these aspects further, let me briefly consider the case of imprinting and artificial insemination.

Imprinting and Cooperative Artificial Insemination

Breeding birds in captivity can be achieved through the making of so-called natural breeding pairs, in which a male bird and a female bird are placed together in an aviary in the hope that they develop a pair bond and show a willingness to mate. Another option is artificial insemination, which can be accomplished either through involuntary or voluntary cooperative insemination. Involuntary insemination may be technically possible, in the sense that involuntarily inseminated birds could produce offspring, but breeders consider it more stressful for the birds. They fear it may lead to a breakdown of good relations with the bird, making it unmanageable. Cooperative insemination is seen to be more effective in comparison with involuntary practices, because the birds are said to be "enjoying" it and are actively involved in the act of courtship (see Fox 1995; Berry 1972). For instance, the female, if stimulated well, will draw in the semen properly, leading to a greater likelihood of egg fertilization (on the practices involved in the artificial insemination of pigs, see Blanchette 2013 and Anneberg et al. 2013).

To breed with birds cooperatively, breeders develop a close bond with male and female birds that have been raised as "social imprints" (Durman-Walters 1994). Social imprinting initially involves raising the nestlings in the breeder's home, where they become part of the everyday routine of the human household. After this period, the breeder needs to continuously maintain a relationship with the imprinted birds so that the strong association with the breeder will eventually lead the birds to sexually identify the breeder as a breeding partner or mate. However, it is important to point out that in the narratives of falconer-breeders, imprinting does not at all present itself as a straightforward process. The falconer-breeders I spoke to often highlighted that the status of a bird as "imprint" was not a fixed, predetermined condition of the bird's identity but rather was subject to the flexible and unpredictable aspects of bird behavior. For instance, birds of prey hand-reared by humans associate strongly with them, but they have also been observed to form pair bonds with other birds if kept together for a long enough period. Birds that have not been imprinted on humans can also develop a pair bond with humans even if other potential bird partners are around. The case of imprinting birds of prey, then, shows that it is difficult to make generalizing claims about imprinting across or even within species.[2]

When breeding with male birds, the breeder takes over the "female" role in courtship (and vice versa). In the case of birds of prey, these ritual roles do not differ significantly between the sexes (Ratcliffe 1980). The event usually involves so-called appeasement gestures in which both partners avert their gaze, bow down their heads, and use different calls that vary in intensity and according to the situation. The moment of copulation itself is very brief. Female birds when aroused and ready will start to "stand," pointing their behind upward so that the male may mount them (breeders may use a hand to put pressure on the bird's back to simulate this part and use a syringe to squirt semen on top of the bird's cloaca). The male bird, on the other hand, after having been courted, will get accustomed to copulating with a particular part of the breeder. In breeding with falcons, this is most often the head of the falconer covered with a rubber hat, which the bird will jump up on and with which it will copulate. The semen will subsequently be collected with a syringe from the hat. Whether the bird will copulate with the designated hat is another question. Some birds are said to prefer to copulate with a falconer's boot, while others prefer a hand or even an external object that can be placed on the floor while the breeder does the calls and bodily movements of courtship. The hats used are also often adjusted to the specific desires of the birds (in terms of size and features such as color) (Wrege and Cade 1977; Weaver and Cade 1991; Platt et al. 2007).

Cooperation in these intimate interspecies encounters presents itself as a multilayered social negotiation, which is ill-described as belonging to either of the extreme sides of "total human control" or interspecies mutuality. Such conceptions would leave us ignorant of the manifold enactments of agency and intimacy that appear to be involved in these interspecies courtship rituals on both sides of the human-bird bond. While the birds are living in captivity and therefore restricted environments and depend on humans for their well-being, they do, within these structures, possess the capacity to act and engage in intersubjective relationships. Intersubjectivity has also been observed to be central for the interactions between humans and nonhuman animals by Despret, for whom this involves a bodily attunement to the other, of learning to affect and be affected by other beings and worlds (2004). Attunement enables "a new articulation of 'with-ness,' an undetermined articulation of 'being with'" (Despret 2004, 131). "Being with" birds of prey in the context described here is, of course, ambiguous, in that it is a situation created by practices of captivity and captive breeding; this chapter thus aims to describe the becoming-with (on agility training, see also Haraway 2008, 206–8) that occurs in these settings, without either justifying or critiquing the settings themselves.

Feeling at Home in the Aviary

The setting that allows for conditions of attunement in breeding practice is the aviary. During my fieldwork, falconer-breeder James showed me around his birds' aviaries: "The most important point [when bringing birds into breeding condition] is that you make your birds feel at home in their aviary. They have to accept it as their own territory and need to be settled and relaxed to allow you to enter their personal space." What becomes interesting, then, are the ways breeding aviaries are designed to resonate with both the breeder's and the bird's sense of place. The structure of the breeding aviary thus reveals itself as a result of a cocreative process pointing to some important relationships between breeders and aviary inhabitants. Furthermore, the success of a captive environment depends on how the birds subsequently make themselves at home within it and come to identify this space as their own. As with other aspects of domestic breeding, when designing these spaces, falconer-breeders usually pointed to parallels to the way birds of prey in the outdoors make themselves at home on cliff ledges or in trees. The aim of the captive environment is to recreate features of these dwelling places while allowing the breeder to approach the birds easily and keep the aviary clean and hygienic. Although the power relations of enclosure cannot be ignored, the design of the aviary, I suggest, should be understood as the outcome of a continuous, improvisational and co-constituted process in which humans and birds are involved (this can be seen in the experiments with aviaries over the years and the fact that breeders are continuously working on them in general, as well as in their efforts to adjust them to the individual idiosyncrasies of particular inhabitants). The aviary furthermore facilitates social relationships among its inhabitants and the breeder, because the management of physical distance is an important part of courtship rituals and must be negotiated between birds and breeders as part of their pair bond (this will be further discussed below; for the use of tools in establishing a communicative relationship between human and bird, see also Anderson et al. 2017, 9, 11–12).

James explained to me that paired birds should have an aviary that allows them privacy and seclusion. The aviaries should provide them with the opportunity to watch and observe what is going on outside while protecting them from the gaze of people and other passing beings (such as dogs, cats, etc.) that might cause them stress and thus potentially hamper breeding success. Because paired birds usually live quite some distance from each other and come to share more intimate space only during the breeding season, these breeding pens must also provide enough room and places to perch for the birds to avoid each other. Breeders, when introducing the birds to a new aviary, also must account for how they will subsequently establish their territory. If the bigger, stronger, and usually more dominant female is introduced to the aviary at the same time as the male, for instance, she is most likely to dominate the male and try to chase him away. This may lead to the male being too intimidated by the female, a situation that will not allow for a breeding partnership to develop. Breeders will usually release the male into the aviary about two weeks before the female arrives, which gives him time to get used to the surroundings and establish himself in his new territory. This is said to "balance" the dominant behavior of the female and thus lead to a more balanced and peaceful relationship between the two in the future. The establishment of a territory or "home," as James called it, therefore involves a certain notion of ownership and control on the side of the falcons, which treat it as their own personal space.

The breeding aviary of imprinted birds, on the other hand, is a very different matter. Because imprints are used to being around people and to the sights and sounds encountered within the human dwelling places, imprint aviaries are not secluded completely from human contact. James constructed his imprint aviary the following way: based in his large back garden, the front of the aviary was open and secured with bars. Inside the aviary in front of this opening was a wooden plank topped with Astroturf to protect the bird's feet on which the bird could perch or walk around overlooking the garden, being able to observe what was going on around him as well as communicate with James or other people. If James was watering his plants, his imprint would sit on the ledge and call to him, thus remaining in regular contact and communication. In the back of the breeding aviary was a "ledge" elevated from the ground so that James could easily engage with the bird for the courting rituals and "ledge display." It consists of a rectangular wooden platform fitted into the corner of the aviary and filled with pea gravel, and it is the place the bird will most likely roost (some birds don't accept these as a roosting place and prefer to stay on the ground or on one of the other perches provided). Early on, when the imprint is still an eyass—a period in which he will live with the breeder in his home—he will from time to time be introduced to the breeding aviary and, ideally, gradually come to identify it with a sense of home. In contrast to the breeding pen for natural pairs, which needs to give the birds enough space to distance themselves from each other, the breeding pen of imprints is constructed in such a way that it will not allow the imprint to evade the falconer when he enters the breeding aviary. James explained that each bird of prey has a "kind of personal space around him." If the breeder enters the pen of the imprint, he immediately enters this personal space and the bird is therefore obliged to engage with him in some way. If the pen were larger, the bird could evade direct contact with the breeder, and it would thus be more difficult and time intensive to involve the bird in the breeding courtship to which I will now turn.

Courtship Rituals

On another visit with James and his wife, Anna, at their home in Wales, Anna invited me to have a cup of tea with her in the kitchen, as "James was busy with the birds." From the open kitchen window, I saw James disappear into one of the aviaries wearing a strange-looking rubber hat. After a while I heard some sounds that gradually cumulated in a loud call (*weeu-weeu*), which was uttered first by James and then by a falcon. Subsequently, James reappeared from the aviary carrying a small glass tube, which he placed in the fridge. During the breeding season, the fridge became his sperm bank, where he kept—next to milk and orange juice—the precious semen donation he received from his tiercel.[3] Anna explained that during the breeding season "James was all bird" and very introverted ("even more than normal"), as he was fully focused on looking after the birds, visiting them between three and four times a day to either collect the semen from the males or inseminate the females.

Here, it should be pointed out that birds of prey reach sexual maturity relatively late, typically for large falcons in their third or fourth year. Breeder and bird therefore enter long-term relationships, and the development of a pair bond needs to be continuously negotiated. Even when birds form a pair bond with each other (or the breeder), they may not necessarily breed together. In fact, when the breeders talk about the bird's ability to breed, they often emphasize that they do not understand mating or successful reproduction as a straightforward instinctual act but rather as something that has to be learned by the birds and that involves skill and responsiveness toward the breeding partner. During the time of courtship, the pair bond needs to be maintained and reinforced. This is done through the performance of specific bodily movements (or courtship displays), the appropriate means of communication (vocal and gestural), and the exchange of food offerings, which are central enactments of affective social relationships. James highlighted the importance of being able to learn these "rules" of courtship, as well as the importance of paying attention to the individual idiosyncrasies, moods, and preferences of the bird "one is courting." In a sense, similar to their breeding birds, breeders need to learn to be responsive to their breeding partners and the specific performance of courtship rituals into which they can indeed enter, despite their different species and bodily shape (for a discussion of affect and mood in the formation of interspecies intimacy, see also Schroer 2018). For the breeders described above, this means that they must commit a considerable amount of time to focusing on breeding activities and that these activities become a dominant force that structures their everyday lives. To make the birds reproduce in captivity, the responsiveness of the breeders and their adjustment to the "rules of bird courtship" are essential criteria for breeding success.

As pointed out above, the bird's sense of personal space is particularly important in structuring social relationships, as well as in the performance of courtship rituals. While breeders can control the conditions in which they interact with imprinted birds (i.e., size of the aviary), they cannot force or coerce the birds into courtship. Birds of prey seem to be much less straightforwardly subjugated to human will than more docile animals, which are more likely to be accepting of social hierarchies. James told of his experience of learning this the hard way at the beginning, having several encounters with enraged hawks that would not let him anywhere close to them in their aviary. "They command the space they live in and if they don't like you there, they will show it." He also highlighted that to enter into courtship with a bird means to show him/her that you come with peaceful attentions. Many of the elements of courtship rituals, such as the greeting call, food offerings, and no direct eye contact, all serve to show an absence of aggressive intent. He explained further: "Very often it means [breeding birds of prey] to accept that the bird might actually be in more control of the situation than you and you just have to play along, see whether you get offered some food or whether you fell out of favour by not having responded well in previous encounters."

While, to some extent, breeders' knowledge derives from the observation of birds and reading descriptions of typical courtship behavior for particular bird species, an important part of their knowledge derives from direct engagement and colearning in relationships with imprinted birds. Although the outdoors courtship rituals of birds include elaborate flying displays in which the male and female engage with each other in the air, the courting rituals in captivity follow similar patterns but are described by breeders as more "condensed versions" of what happens in the wild. To observe birds in the captive environment, breeders therefore have to gradually learn to interpret their calls and bodily movements, and, as James pointed out, the differences can be quite nuanced and need to be interpreted within a wider context and, importantly, in relation to the individual bird.

Bird Desire

Breeding with birds, as we have just seen, is a multilayered task that cannot easily be compartmentalized into successive steps. In this sense, breeding birds of prey differs from other domestic-breeding practices that have been described in which human-animal intimacies are involved. In large-scale pig breeding in industrialized meat production, for instance, Blanchette describes the intimate human labor and skill that goes into stimulating sows to draw up the semen (Blanchette 2013; see also Anneberg et al. 2013). Similar to the above example, Blanchette speaks of "pig desires" that in this context are put into production routines to provide the best conditions of conceptions and to ultimately increase the number of piglets conceived. The increasing standardization and automation of processes in the industry, however, has led to the development of technologies that mimic humans mimicking of animal behavior, such as the "portable AI assist," a belt that reduces the need for human stimulation of sows in preparation for insemination by physically holding the semen in (Anneberg et al. 2013). For birds, on the other hand, as James explained above, it is central to form a long-standing personal bond and to be responsive to each individual bird's needs and desires. "This is something you cannot standardise," he explained. "Each bird will have different things that stimulate him or her, different times of the day in which they will be more likely to be in the mood than others. Some have to be courted for more than 30 minutes, others will be ready to go as soon as they see me approach their aviary." In short: what works for one bird might not do for another.

Courting rituals are, on the one hand, a performance of falcon "lust." But, on the other, they are also a performance of a very specific relationship, namely, that between breeding pairs that follow the demands of both partners. When engaging with birds on a daily basis, breeders experience the birds as beings with particular desires, and, for people who want to breed them, this is an essential point of departure that effects *everything*, including the design and building of the breeding aviaries, the use of communication, bodily gestures, and specific calls, the choice of food, regulation of temperature and humidity, and the aesthetics of tools such as the breeding hat. Bird desire here can be understood both in terms of a sense of sexual longing and lust and in terms of having (at times quite high) demands and expectations of their breeding partner. Speaking of bird desire points to the importance of understanding these birds as actively involved in the negotiation of relationships. They are not limited to the binary choice of merely accepting or resisting relationship; rather, they are able to shape the relations in which they engage in a variety of ways. As the case of "imprinted" birds shows, this desire is also not limited to members of the same taxonomic species but is a result of the development of the individual bird. The notion of desire also implicates a certain sense of interspecies moral conduct of how things should or should not be done. For breeders, it becomes important to respond to these desires, which turns their acts of domestic breeding into a highly intimate process that necessitates coresponsive participation and sensitivity on the part of both the humans and the birds involved. It is not a mechanistic act of production. Reproduction in the context of the captive breeding establishment, then, is constituted and achieved through carefully choreographed collaborative acts of commitment and intimacy among birds and humans who perform different roles depending on their form of participation in the breeding complex. The narratives of falconers who breed birds are filled with rich information about these elaborate means of communication, as well as social negotiation that is necessary when breeding. They present an image of a process that is a far cry from the idea of "complete mastery" by humans via so-called artificial insemination. Birds have to be in the mood; they have to be charmed and made ready for mating. Otherwise, nothing happens. The birds, then, are clearly not just passive beings acted upon by humans, nor are they beings that are easily controlled or dominated in interspecies intimate encounters.

Interspecies Intimacy

When considering intimate interspecies engagements in practices of domestication, it seems important to differentiate between types of possible intimacies between humans and other critters. The sexual intimacies and cooperative reproductive activity between humans and birds differ significantly from both relations between humans and pets—often portrayed as asexual relationships—and the relations between people and working animals (e.g., see different case studies on human-animal intimacies in Knight 2005). While breeders state that they do not themselves get sexually aroused during these intimate encounters, they still need to perform in a responsive and authentic way to "please" the birds, if they are to ensure their cooperation. Second, the relationships and intimacies between humans and birds are not determined or fixed as a single possible configuration. The form of sexual intimacy that becomes relevant primarily during the breeding season is not the only possible form of social association. Breeding birds, for instance, are very often also falconry birds, trained to cooperate in hunting. Some falconer-breeders argue that flying them through the use of falconry methods and letting them hunt contributes to their overall well-being, as well as to the development and maturation of the bird. According to the breeders, this formation of a social bond between human and bird will then enhance the development of a breeding partnership once the bird reaches sexual maturity.

Speaking of intimacy and sociality in practices of domestication is, of course, nothing new. Arguments toward understanding human relationships with domesticates in these terms have often been formulated in counter to Ingold's distinction between trust (as characteristic of the relationships between hunter-gatherers and prey) and domination (as characteristic of those between humans and domesticates) (Ingold 1994; for examples of these counterpositions, see Knight 2005, 2012; Oma 2010). Nonetheless, rather than destabilizing these contrasting dualities, there is a tendency in these approaches to reinscribe dualisms (e.g., of wild and domestic) through reversing the argument and assuming that intimacy and sociality are part of the domestic sphere of human-animal relations but cannot be found in hunters' relationships to wild animals (see also Hill 2013 for a further critique). When doing ethnographic research in the midst of domestication practices, these "either-or" scenarios seem simplistic and detached from actual cross-species encounters in particular situations. In the human-bird relationships that I have described, intimacy would not exist were it not for dominant and controlling acts by humans, but—at the same time—that intimacy is not devoid of bird agency.

Difference and similarity in these intimate encounters are framed not only by species boundaries but also by one's ability or inability to communicate and respond as a breeding partner. This interspecies responsiveness shows the limits of the assumption that intimacy in human-animal relationships is most readily found between humans and other mammals, due to their phylogenic relatedness (for an example of this reasoning, see Knight 2005, 11). In other words, the conditions for interspecies sociality should not be understood as being based on a priori taxonomic belonging, "mental capacities," or "cognition." As Anna Tsing recently argued, to assume so limits our curiosity and the possibilities of discovering "more-than-human sociality" in places and relationships where we would otherwise not look (Tsing 2013). Focusing on sociality as emergent and unconstrained by species barriers therefore requires us to decenter thinking in terms of "the Human" and "the Animal" in domestication and instead pay attention to the practices through which particular living beings develop in relation to each other and, to paraphrase Despret (2008), accomplish subjectivities depending on the situations in which they interact.

Conclusion

Drawing on ethnographic fieldwork with falconer-breeders and birds of prey, this chapter has argued that to understand domestication in relation to captive breeding of birds of prey, one must acknowledge nonhumans as active participants in social relationships, through which both human and nonhuman beings learn to be responsive to each other. As we know, control over reproduction is a prominent element in traditional definitions of domestication. While it is clear that breeding takes place within captive environments—a setting in which humans are making use of their knowledge of imprinting to enable a particular bond to develop—the birds' agency is not irrelevant. We can observe how in falconer breeding knowledge of social imprinting, the thread of species extinction, new legislation (e.g., pertaining to the capture of chicks in the wild), and the emerging setting of artificial insemination have enabled novel ways of interacting that transform both bird and breeder (Despret 2004, 8). This involves interspecies relations that are continuously negotiated in a situation of varying *intensities* and with a range of affective relations, encompassing dominating, aggressive, or fearful responses as much as affectionate, appeasing, and trusting engagements.

Such observations lead to the realization that domestication should not be understood as a universal, monolithic concept but rather as a dynamic set of interspecies practices that are specific, temporal, and situated. Describing the manifold relationships involved in contexts of domestication allows us to learn about forms of more-than-human sociality as they are enacted in the everyday life of humans and other living beings.

NOTES

1. This chapter is based on initial research into captive breeding of birds of prey, conducted as part of the ERC-funded project Arctic Domus at the Department of Anthropology, University of Aberdeen. It also builds upon my doctoral research, in which I explored the relationships between humans and birds of prey in falconry. This research was based mainly on ethnographic fieldwork in Britain as well as smaller field trips to Germany and Italy (cofunded through the International Rotary Foundation, Sutasoma/Radcliffe Brown Trust of the Royal Anthropological Institute, the Principal's Excellence Fund University of Aberdeen, Falconry Heritage Trust, and Deutscher Falkenorden).

2. In the light of such ethnographic insights, as well as through the lens of current concerns with better understanding more-than-human sociality, it seems worthwhile to revisit and potentially critique early ethological studies on the imprinting of birds (see, for instance, Lorenz 1937, 1965; Hess 1973) that still are used for understanding processes of imprinting today. These portray imprinting as a mechanistic, determining, and irreversible event in the early life of a living being, firmly based in the conventions of ethological scientific writing at the time. For reasons of space, this cannot be further explored in this chapter.

3. A falconry term for a male falcon.

REFERENCES

Anderson, David, Jan P. L. Loovers, Sara A. Schroer, and Rob P. Wishart. 2017. "Architectures of Domestication: On Emplacing Human-Animal Relationships in the North." *Journal of the Royal Anthropological Institute* 23: 398–418.

Anneberg, Inger, Mette Vaarst, and Nils Bubandt. 2013. "Pigs and Profits: Hybrids of Animals, Technology and Humans in Danish Industrialised Pig Farming." *Social Anthropology* 21, no. 4: 524–59.

Berry, Robert B. 1972. "Reproduction by Artificial Insemination by Captive American Goshawks." *Journal of Wildlife Management* 36, no. 4: 1283–88.
Blanchette, Alexander D. 2013. "Conceiving Porkopolis: The Production of Life on the American 'Factory' Farm." PhD thesis, University of Chicago.
Bökönyi, Sandor. 1969. "Archaeological Problems and Methods of Recognising Animal Domestication." In *The Domestication and Exploitation of Plants and Animals*, ed. Peter Ucko and G. W. Dimbleby, 219–29. London: Duckworth.
Bökönyi, Sandor. 1989. "Definitions of Animal Domestication." In *The Walking Larder: Patterns of Domestication, Pastoralism and Predation*, ed. Juliette Clutton-Brock, 22–27. London: Unwin Hyman.
Cassidy, Rebecca. 2002. *The Sport of Kings: Kinship, Class and Thoroughbred Breeding in Newmarket.* Cambridge: Cambridge University Press.
Cassidy, Rebecca, and Molly Mullin, eds. 2007. *Where the Wild Things Are Now: Domestication Reconsidered.* Oxford: Berg.
Clutton-Brock, Juliette. 2012. *Animals as Domesticates: A Worldview through History.* East Lansing: Michigan State University Press.
Despret, Vinciane. 2004. "The Body We Care For: Figures of Anthropo-zoo-genesis." *Body and Society* 10, no. 2/3: 113–34.
Despret, Vinciane. 2008. "The Becomings of Subjectivity in Animal Worlds." *Subjectivity* 23: 123–39.
Durman-Walters, Diane. 1994. *The Modern Falconer: Training, Hawking, & Breeding.* Wykey: Swan Hill.
Fox, Nick. 1995. *Understanding the Bird of Prey.* Surrey: Hancock House.
Haraway, Donna J. 2008. *When Species Meet.* Minneapolis: University of Minnesota Press.
Hess, Eric H. 1973. *Imprinting.* New York: Halsted.
Hill, Erica. 2013. "Archaeology and Animal Persons: Towards a Prehistory of Human-Animal Relations." *Environment and Society: Advances in Research* 4: 117–36.
Ingold, Tim. 1994. "From Trust to Domination: An Alternative History of Human-Animal Relations." In *Animals and Human Society: Changing Perspectives*, ed. Audrey Manning and James Serpell. London: Routledge.
Ingold, Tim. 1995. *Key Debates in Anthropology.* London: Routledge.
Ingold, Tim. 2000. *The Perception of the Environment: Essays on Livelihood, Dwelling and Skill.* London: Routledge.
Ingold, Tim. 2011. *Being Alive: Essays on Movement, Knowledge and Description.* London: Routledge.
Knight, John. 2012. "The Anonymity of the Hunt: A Critique of Hunting as Sharing." *Current Anthropology* 53, no. 3: 334–55.
Knight, John, ed. 2005. *Animals in Person: Cultural Perspectives on Human-Animal Intimacies.* Oxford: Berg.
Long, Nicholas J., and Henrietta L. Moore. 2012. "Sociality Revisited: Setting a New Agenda." *Cambridge Anthropology* 30, no. 1: 40–47.
Lorenz, Konrad. 1935. "Der Kumpan in der Umwelt des Vogels: Der Artgenosse als auslösendes Moment sozialer Verhaltensweisen." *Journal für Ornithologie* 83: 137–213, 289–413.

Lorenz, Konrad. 1965. *Evolution and Modification of Behaviour.* Chicago: University of Chicago Press.

Oma, Kristin A. 2010. "Between Trust and Domination: Social Contracts between Humans and Animals." *World Archaeology* 42, no. 2: 175–87.

Platt, Joseph B., David M. Bird, and Lina Bardo. 2007. "Captive Breeding." In *Raptor Research and Management Techniques*, ed. David M. Bird et al. Surrey: Hancock House.

Ratcliffe, Derek A. 1980. *The Peregrine Falcon.* Vermillion, SD: Buteo.

Roots, Clive. 2007. *Domestication.* London: Greenwood.

Sax, Boria. 2014. "Jocelyne Porcher, an Introduction." *Humanimalia* 6, no. 1: 1–4.

Schroer, Sara A. 2018. "'A Feeling for Birds': Tuning into More-Than-Human Atmospheres." In *Exploring Atmospheres Ethnographically*, eds. Sara A. Schroer and Susanne B. Schmitt. London: Routledge.

Schroer, Sara A. 2015. "'On the Wing': Exploring Human-Bird Relationships in Falconry Practice." PhD thesis, University of Aberdeen.

Spalding, Douglas A. 1872. "On Instinct." *Nature* 6: 485–86.

Stark, J. Matthias, and Robert E. Ricklefs, eds. 1998. *Avian Growth and Development: Evolution within the Altricial-Precocial Spectrum.* Oxford: Oxford University Press.

Tsing, Anna. 2013. "More-Than-Human Sociality: A Call for a Critical Description." In *Anthropology and Nature*, ed. K. Hastrup. London: Routledge.

Weaver, James D., and Tom J. Cade. 1991. *Falcon Propagation: A Manual on Captive Breeding.* Ithaca: Peregrine Fund.

Wrege, Peter H., and Tom. J. Cade. 1977. "Courtship Behaviour of Large Falcons in Captivity." *Raptor Research* 11, no. 1/2: 1–46.

2

PIGS AND SPIRITS IN IFUGAO

A Cosmological Decentering of Domestication · *Jon Henrik Ziegler Remme*

It was early morning, and eight men and I had gathered under the pillared house of Mahhig, one of my neighbors in a village in Ifugao, a highland province of Northern Luzon, the Philippines. Wigan, a male cousin of Mahhig, was about to get married to Appeng, a girl in the neighboring village, and we had been summoned to help catch Mahhig's pig and carry it there as part of the *madāwat,* the series of bride wealth prestations that Wigan was giving to Appeng's parents. We soon realized, however, that the pig would be no easy catch. Several times we spread around in the courtyard, tried to encircle the large, black hog, but as soon as we came close, the pig rammed right toward us. To avoid getting hurt, we had to let it slip between us and run out of the yard and into the forest. After a good hour of thus chasing it between the forest and the yard, we eventually cornered the pig under the house and threw a net over it. Two of the men tied its legs to a bamboo pole and raised the pig onto their shoulders.

After a couple of hours' walk through the maze of trails running along the rice terrace walls, passing by steep swidden fields and through thick forest, we approached Appeng's village. Before entering, we made sure everyone carried a piece of ginger, since the smell of it would protect us against the spirits of this village and against *pāliw*, a form of envy-generated witchcraft. We carried on and eventually arrived at Wigan's future parents-in-law's house. We rested there for a while but left before dark, when the ritual experts would come to the house to commence the sacrificial rituals in which the pig would be offered to the spirits.

The hunt-like character of the pig chase that morning and its implication for human-spirit relations indicate the main points I want to discuss in this article, namely, how Ifugao relations between humans and pigs, domestic and otherwise, challenge conventional narratives of domestication and spur us to think domestication otherwise. As I will show, Ifugao domestication intervenes in those narratives in various ways: by showing that various forms of domestication often coexist, that these domestication practices occur within a relational field that is more extensive than is often assumed, that the distinction between domestic and nondomestic animals are characterized by blurred boundaries, that domestic animals are only partially contained and controlled by humans, and that domestication practices are part of and consequently implicated in wider sociopolitical processes.

Conventional narratives of domestication often portray domestication as an epochal change in human-animal relations that occurred when humans began to modify or control the breeding of animals (Bökönyi 1989), incorporated them into the social structure of the human group, made them objects of ownership (Ducos 1989), or alternatively, changed their relations with them from being based on trust to being characterized by domination (Ingold 2000).[1] These transformations in human-animal relations are also often held to imply the unilinear and irreversible transition of humans from hunters to pastoralists (Cassidy 2007, 8).

Although there is no need to underplay the historical significance of the first attempts at domestication that took place about ten to twelve thousand years ago, there is a tendency in such "epochal approaches" to overstate the differences between hunting-and-gathering and agricultural societies and consequently to undervalue the significant fact that different forms of domestication and different forms of human-animal relations coexist and have continued to do so even after domestication became the dominant form of human-animal relations. There is, for instance, ample ethnographic evidence to support the claim that the domestication of a species does not necessarily encompass all specimens of that species and that relations of domestication can coexist with other forms of relations between humans and that species. Domestication of reindeer does, for instance, not exclude hunting of reindeer (Vitebsky 2005, 25, 377; Willerslev, Vitebsky, and Alekseyev 2014, 6, 12), and the current thriving industry of salmon farming (Lien 2015) does not exclude the practices of fishing for wild salmon. Indeed, the scale and importance of global fisheries today clearly indicate that hunting and trapping as modes of relating to animals have far from become sidelined by the introduction of domestication but coexists with it, and in many cases these forms support and draw upon each other in terms of knowledge production, infrastructure, and consumption. Correspondingly, the ethnography from Ifugao shows clearly that these different forms of domestication and human-animal relations not only coexist alongside each other but interrelate in specific ways that incite us to think domestication in terms of landscape transformations, as Heather Anne Swanson argues in this volume, or as ongoing transformations of multispecies assemblages.

Hence, to fully grasp the dynamic interrelations of these coexisting forms of human-animal relations, we need to decenter the epochal approaches to domestication and rather approach domestication as an ongoing and shifting multiplicity of forms of relations between humans, animals, and plants (Oma 2010).

The interrelations of these coexisting forms of domestication invite us thus to extend the relational field of domestication beyond the relations between humans and a particular species to a wider more-than-human sociality (Tsing 2013). However, to fully understand how these interrelations operate in Ifugao, we need to account for what we might call the cosmological frame (Willerslev, Vitebsky, and Alekseyev 2014; Abramson and Holbraad 2014) in which these domestication practices occur. By "cosmological frame" I refer to the ideas and practices related to how the relations between humans and all the other living beings populating the Ifugao world—which includes animals and plants but also, and quite significantly for this account, the variety of other-than-human beings (Cadena 2010) like ancestors—place specific *pinādeng* spirits and a host of others (see Barton 1946). Relations between humans and spirits are of vital concern for human health and well-being in Ifugao, as well as for the outcomes of agricultural efforts. Pigs, domestic or otherwise, figure centrally in the enactment of these relations, and the various plant and animal domestication practices are here always embedded in multispecies assemblages that include spirits in one way or another. The relations that make up this cosmology are far from stable, however. Spirits are occasionally ignored, at other times they become highly actualized and present, and humans and animals may be partially, temporarily. or more permanently transformed into spirits themselves. Social relations may thus extend far beyond the human domain, and the more-than-human sociality thus extends and contracts continuously (Remme 2017).

When I argue for paying attention to the cosmological frame in which domestication practices take place, I should underscore that I do not intend to set up a clear distinction between a Western and an Ifugao cosmology. The boundaries of Ifugao cosmology, and Western cosmology for that matter, are far too blurred and contextually shifting to allow such a contrastive move. I emphasize the cosmological frame here because I find that taking seriously my Ifugao interlocutors' claims about the nonhuman beings with which they share their world and lives has profound consequences for what domestication becomes. It challenges, for instance, the often taken for granted distinction between domestic and nondomestic animals and the processes through which transformations of these are done. While my Ifugao friends tended to make a clear distinction between different kinds of pigs, for instance, between what they translated to me as domestic and wild pigs, they also acknowledged that these categories were specific to their point of view as humans. They knew that those pigs they saw as wild pigs were also something other-than-wild, as they were the domestic pigs of the *pinădeng* spirits. They respected this double quality of pigs and acted upon it when they went hunting or sacrificed pigs in rituals. As I will show, the cosmological framing of domestication practices thus blurs the boundary between domestic and nondomestic animals in a way that unsettles distinctions that are so central to conventional narratives of domestication (see Lien, this volume; see also introduction). Indeed, the cosmological perspective challenges us to rethink the very ontology or mode of being of pigs. Pigs in their various modes of being contain an excess imminent to their own existence (Cadena 2014; Povinelli 2014), a nonactualized inversion that can occasionally become actualized, which, as I will show, occurs during hunting and at sacrificial rituals. Furthermore, as indicated in the opening vignette, domestic pigs in Ifugao are kept close to the house but are nevertheless quite "slippery" (Law and Lien 2013), in the sense that they often traverse categorical distinctions like forest and house yard and for much of the day actually occupy the space between them (Fuentes 2007; Lien 2007). I will show that the relations between humans and spirits are crucial for this slipperiness and that the emphasis on partial containment in Ifugao pig husbandry challenges notions of control, containment, and human mastery over animals—other components central in conventional domestication narratives.[2] The partial containment of domestic pigs is imperative for making the pigs suitable as sacrificial offerings, which is the main purpose of domestic pigs here, but it is also crucial for the health and well-being of humans themselves. Proper landscape and husbandry practices are conducive to this, as is the occasional sacrifice of the pigs to the spirits. The spirits will react with vengeance, however, if these practices are not done according to their ways. Domestication is therefore far from representing human mastery. It is rather a quite risky business for humans, a matter of life and death.

The cosmological approach I take here does not entail that Ifugao domestication practices are not also embedded in wider sociopolitical processes. The villagers currently experience an increasing involvement in a monetary economy and in national political processes, both of which contribute to transforming the shape of the coexisting assemblage of domestication forms. New forms of pig husbandry are emerging, driven partly by commercial enterprises and partly by governmental efforts at protecting and developing the pig stock endemic to the Philippines. And religious conversions to Pentecostalism transform both the villagers' landscape transformations and their involvement in human-pig-spirit relations.

The Pigs and the Landscape

The pig we chased in and out of the yard that morning had lived in and around Mahhig's house for its entire life and had become part of Mahhig's family. When it was born, Mahhig and his wife, Bugan, had kept its mother and her piglets in a pigsty under the house. Just as they did when their own children were born, when the piglets were born, they had stuck a grass tied into a particular *pūdung* knot onto the gate leading into the yard. This would signify to others, including the spirits, that the house contained newborns and that they should not enter the house, to avoid bringing with them potential harmful influence, for instance, the ominous influence of a bad omen. The knot also protected against witchcraft. After two weeks of containing them within the pigsty, Mahhig removed the *pūdung* and let the sow and her piglets roam free on their own. The mother led the way, her piglets tripping around her, sniffing their way along the village paths. The sow usually wore a *bahāngal*, a form of wooden necklace tied to one of her feet. This restricted her gait and thus impeded her mobility just about enough to let her roam around with her piglets without going too far into the forest. In the evening, Mahhig always made sure the pigs returned, so that he could lock them into the pigpen placed right behind the house.

The mobility enjoyed during daytime allowed the pigs to move across spatial divides, such as the forest areas, which are the prime area of the *pinādeng* spirits, and the house, *baluy*. The *baluy* has a central role in establishing and mediating relations, both between humans and between humans and animals. When a man and a woman marry, they are referred to as *himbaluy*, and their children, nephews, and nieces are their *imbaluy*, both meaning "of the house." The association between the family and their pigs are also mediated through the house. They are kept within the house when born and always return to the house at night. When they are later sacrificed, they are also placed at particular parts of the house and are sometimes moved around the house to show, as one of my informants put it, "that it is sacrificed instead of the house." The mythical explanation for the use of pigs as sacrificial animals also indicates this, as it relates how the spirits instructed the humans to substitute pigs for their *imbaluy*, their offspring. Although the pigs are thus closely associated with the *baluy* and become part of the family living there, they are allowed to move between the *baluy* and the forest. The containment within the confines of the *baluy* is thus partial and must actually be so for the pigs to later become suitable as sacrificial animals.

During the day, Mahhig and Bugan's pigs fed on anything edible they came across, but in the afternoons, they also received fodder. This came from the *ūma*, the swidden field where Bugan worked most days, weeding, planting, and harvesting carrots, ginger, taro, and most importantly sweet potatoes. Most of the sweet potatoes they ate themselves, but those they found unfit to consume were given to the pigs along with other edible weeds brought down from the *ūma*.

The pigs were also fed products from the *payaw*, the irrigated fields that the couple own. Most of the mountainsides that encapsulate the village have been carved into terraces, and an extensive system of stone-built channels leads the water down from sources in the watershed forests on top of the mountains. These terraces are part of the heritage a person receives upon marriage, although far from all receive them, since it is usually the first-born child who inherits his or her parents' terraces. The *payaw* are often cultivated individually, and it is men who build and repair the stone walls and plow either by hand with shovels or with a water buffalo, for those who can afford one. Most of the *payaw* are used for cultivating wet rice, some of which is of a particularly glutinous variant used for making rice wine, and often the terraces are referred to simply as rice terraces. However, this translation betrays the nature of the *payaw* as multispecies assemblages. Besides being used for wet rice, the *payaw* are a source for a variety of other foodstuffs. Vegetables and fruits like pineapple are grown on the stone and mud walls, and the ponds contain large amounts of snails and mud fishes, some benefiting rice cultivation, others obstructing it.

Maintaining these terraces in proper condition is vital for life in Ifugao. The impressive sight of the extensive system of terraces that covers the entire valley betrays also, in a way, their very fragile nature. The terraces are quite prone to disintegrating as soon as they are not kept in proper condition, and they therefore require continual maintenance. Much of the daily work in the *payaw* involves warding off potential harmful forces. For instance, the harmful snails and worms that abound in the ponds must be removed before rice seedlings can be planted. Some of these snails are quite easily picked up by hand while wading through the mud. Others require hiking into the mountains to gather plants that are ritually induced to emit a smell that repels the worms called *lāgab*. These worms are prone to creating holes in the terrace walls, which in heavy rain fill with water that eventually bursts through the walls, ruining the terrace and usually those below, as well (Remme 2014a). Rats too do damage to the terraces, and weeding and keeping the forest from growing too close to the terraces will help them stay away. The weeding and daily maintenance of the terraces are done primarily by women, who bring weeds picked from the stone walls and plants removed from the border between the terraces and the forest, as well as other surplus edibles, back home to feed the pigs. Keeping the terraced landscape of the valley in proper condition is thereby directly implicated in pig husbandry. That the products of this landscape maintenance are consumed by both humans and pigs—often in complementary ways in the sense that humans eat the underground portions of sweet potatoes and taro, while the pigs eat the vines and upper portions of these plants—contributes to the establishment of family relations between humans and pigs.

The implication of pig husbandry and landscape maintenance goes both ways, however. For although the potential destructive forces of rats, snails, and water can be warded off through maintaining the terraces, this would be all but futile if the pigs were not from time to time sacrificed to the spirits. The Ifugao are known for their complex spirit world (see, e.g., Barton 1946). For the sake of brevity, I will refer to this world as consisting of an array of other-than-human beings that include, among others, ancestors, mythical characters, and *pinādeng* spirits that reside in and around the village. Relations with spirits are both potentially harmful and potentially beneficial for humans. Such relations are the source of life and reproduction, in terms of both health and procreation and agricultural yields, but if these relations are distorted in one way or another, the spirits will cause illness and death.

The spirits also directly influence the outcomes of humans' agricultural efforts. They may send rain, draught, or thunderstorms if they are dissatisfied, but if humans relate to them as they should, the spirits will bestow rice in abundance and provide sufficient water for irrigation. Maintaining the terraces is part of enacting well-functioning relations between humans and spirits, as it is held that conducting rice cultivation in the traditional manner is a sign of respect for the spirits, particularly for the ancestors but also for those mythical characters who built the terraces in the first place. Keeping rats and insects away, weeding, and removing snails and worms thus involve more than human-animal relations. The landscaping practices of maintaining the rice terraces is thus a multispecies assemblage that also includes these spirits. Pigs, too, are part of these assemblages, as feeding them waste products of terrace maintenance and swidden cultivation contributes to keeping the multispecies terrace assemblages in a stable condition.

Hence, continuing "the ways of the ancestors" in terrace maintenance is crucial for fruitful human-spirit relations. This goes for pig husbandry, as well. Keeping pigs as part of the family, feeding them food from *payaw* and *ūma*, allowing them to run relatively freely during the day and containing them within the yard at night: all these practices are prescribed by the ancestors as proper forms of pig husbandry. However, pig husbandry is involved in the maintenance of the landscape in another way as well, since it is through this form of pig domestication that these pigs become suitable sacrificial animals. It is through the occasional sacrifice of pigs to the spirits that the landscape can be maintained and can continue to be a source for human life and reproduction. In the following section, I will discuss how the domestication practices that make these pigs sacrificial animals must be understood in relation to other forms of human-pig relations, as well as in relation to human-spirit relations.

Spirit Pigs and Human Pigs

The pig that we chased in and out of Mahhig's yard that morning was of the kind referred to in Ifugao as *bābuy*. The *bābuy* type of pig is differentiated from two other major categories of pigs: the wild pig, *lāman*, which I will focus on here, and the imported pig, *bangyādu*, which I will return to in the next section.

Much of the landscape that surrounds the village is developed into irrigated terraces and swidden gardens, and other areas are covered by forest. Forested areas are divided into *pinūgu*, privately owned forest plots, and *inalāhan*, the public free-for-all forest. Many of these forest areas cover mountaintops and ridges and serve as vital watersheds for the irrigated terraces. They are also areas where wild animals, particularly *lāman*, can be found, although these are claimed to be fewer and far between nowadays. There may be several reasons for this, but one of the most evident seems to be that the forest area diminished considerably between 1990 and 2008, partly driven by the increased demands by the tourist industry in Banaue for logs for carved wood souvenirs. This has led not only to an increasingly unsteady water supply for irrigation but also to the migration of wild pigs to safer territories, perhaps further north where there has been less strain on the forest.

Hunting of *lāman* occasionally takes place anyway, and in these cases the hunting expeditions have to be preceded by rituals in which the hunters are given acceptance by the owners of the *lāman*, the *pinādeng* spirits, to trap down and kill the animals. While the *inalāhan* (public forest) is formally owned by the Philippine state, the *pinādeng* are its spiritual owners. They own therefore also the *lāman*, as well as a range of other animals that humans may encounter in the forest. Although the ownership distribution of animals is relatively clearly delineated, the relations between spirits, wild animals, and humans are complicated by the perspectival differences between humans and spirits. While humans and spirits may be said to inhabit the same landscape, they also differ from each other in terms of how they see things in the world. Humans and spirits share a form of life force called *lennāwa*, but they have different bodies, *odol*. It is the integration of *lennāwa* in a human *odol* that makes humans see the world in their particularly human way. Spirits have a different *odol*, which is mostly invisible to humans, and this causes them to see the world differently; although from their point of view, the way they see the world is similar to how humans see it.

There are thus many parallels between Ifugao cosmologies and Amerindian perspectivism (Viveiros de Castro 1998), as well as with Chewong "relativity in perception" (Howell 1989). These perspectival differences imply that what humans see as wild animals—*lāman* and rats—are for the spirits something different. It is of particular interest here that these (for human) wild animals are the domestic animals of the *pinādeng*. *Lāman* are thus the *bābuy* of spirits, and rats are their domestic chickens.

When the villagers hunt, then, they go into the domain of the *pinādeng* to fetch something that is owned by spirits. There are several ways to do that, both alone and in groups. Shotguns are available now, but before these were introduced, people hunted with spears. The most common form of hunting in the area seems to have been to dig a pit, sometimes with bamboo sticks at the bottom, and for a group of hunters to encircle the pig and chase it into the pit.

However, none of these methods would be successful, people claimed, if the hunt had not been preceded by a ritual to get permission from the *pinādeng* to take the animal. As Ramon once claimed, the spear of even the best hunter would miss the target without the *pinādeng*'s approval.[3] It is also crucial for the hunting party to listen to the calling of a particular bird called *īdaw* while they commence their hunting expedition. If the twittering of the bird is of an ominous kind, they should return to the village and postpone the hunting for another day. When they eventually capture an animal, the initiator of the hunt will receive the skull of the animal and the major share of the meat, while the other hunters share the rest. After a period of hunting success, the hunter arranges a sacrifice to reciprocate for the animals given to him by the *pinādeng*, and the hunter should eventually bring the skulls and bones of the hunted animals back to the forest and thus "return" the animals to their domain, that is, to the spirit owners.

The main point here is that the differentiation between wild pigs and domestic pigs is far from straightforward, since Ifugao cosmology deems that these categories are a matter of perspective. For humans, wild pigs are wild pigs, but, for hunting to be successful, they must show that they recognize that for the *pinādeng* they are domestic pigs. Thus, in the hunting permission rituals and the subsequent hunt, humans enact the animals as both wild pigs and domestic pigs simultaneously. To catch them, they must build pits and traps and chase them through the forest, but, at the same time, they must acknowledge that ultimately it is the *pinādeng* who give them the pigs.[4] Hence, during hunting both the domestic and the wild mode of being of *lāman* become actualized through a double enactment. The cosmological frame in which these forms of human-animal relations take place thus challenges the very ontological ground on which conventional domestication narratives build, since the pigs are here temporarily enacted as both nondomestic and domestic at one and the same time.

The Risk of Domestication

This double enactment of pigs is, however, not restricted to hunting practices but appears also in sacrificial rituals. And it is for these rituals that the *bābuy*, humans' domestic pigs, have been bred and fed. The *bābuy* are killed and eaten solely in rituals and always for the purpose of drawing upon the forces of the spirits. Sacrificing pigs is done frequently in the village and is crucial for the negotiation of relations between humans and spirits. Sacrifices are done for a variety of purposes: for reciprocating successful hunting, for ensuring good harvest, and for healing and blessing purposes. All such sacrificial rituals follow the same pattern, with spirits invited to the house by ritual experts to receive offerings of pigs. It is of absolute importance for the successful effect of these rituals that the pigs offered are *bābuy*, that is, humans' domestic pigs. *Lāman* are unsuitable as sacrificial animals.

So, although we left the neighboring village before Wigan's future parents-in-law killed the pig we had brought, we knew that later that night they would sacrifice it in a ritual. We did not stay long enough to observe it, but when we arrived back at our village, we heard news of another more or less identical ritual that would be held the following day. When Ramon and I arrived early in the morning the next day, we climbed the ladder into the house, and inside we found three men sitting on the wooden floor with a jar of rice wine between them. These were the *mumbā'i*, ritual experts, and they had been there since midnight, invoking the spirits and asking them to come to take part in the feast. Duntugan, the man who owned the house, had been ill lately and kept dreaming that his deceased father was angry with him for not having visited often enough when he was dying. The *mumbā'i* had instructed him to get hold of three *bābuy*, and the large animals now lay in the courtyard with their feet tied together ready to be killed and given to the spirits.

The illness Duntugan experienced was explained as the momentary separation of his life force, *lennāwa*, from his body, *odol*. The *mumbā'i* claimed that Duntugan's *lennāwa* had been called into the world of the spirits, and the only way to retrieve it and thus recover Duntugan's health was to give the spirits pigs so that they would release the *lennāwa* and return it to Duntugan's *odol*. As mentioned above, the perspectival differences between humans and spirits rely on the integration of *lennāwa* with different kinds of *odol*. However, the different *lennāwa-odol* integrations are by no means fixed and are susceptible to dissolution. The consequence of such separations becoming prolonged can be rather dramatic, as it can lead to the metamorphosis of the human into a spirit. Hence, now that Duntugan's *lennāwa* was separated from his *odol*, he risked dying and thus transforming into an ancestor spirit himself. To remain human, Duntugan had to reintegrate his *lennāwa* into his *odol*, and that could be done only through the giving of *bābuy* to the spirits.

This was the aim, then, of Duntugan's ritual. The *mumbā'i* called the spirits, and they were held to come to be present inside the house. As more and more visitors arrived at the house, the *mumbā'i* moved down the house ladder and sat down on the stone-covered ground beneath the house door. There, they commenced invoking the spirits.

When the spirits arrived down at the ground area, the *mumbā'i* became possessed by them. In their possessed state, the *mumbā'i* dressed up in various paraphernalia: a ritual backpack, a headdress, and a spear, all of which are associated with hunting or headhunting. Dressed in various combinations of these items, the possessed *mumbā'i* danced over toward the pigs that lay in the yard, always with the spears aiming for the animals. This part of the ritual is always quite tense, partly because the otherwise rather distant and inaccessible spirits are held to be manifestly present in the yard but also because the ritual's efficacy rests to a large extent on the success of the following rite. One of the *mumbā'is* took a rooster and placed it carefully on the back of one of the pigs. The rooster lay still and did not fly off the pig until the *mumbā'i* hit it lightly with a hunting spear. Then the ritual could continue, and the pigs could be killed. This small but very significant rite is called *umīdaw*, which refers to the *īdaw* omen bird mentioned above. Just as the hunt relies on a portentous calling from the *īdaw* bird, the sacrifice cannot continue without an assuring sign from the rooster-cum-*īdaw*. In this part of Ifugao, the pigs are eventually killed by knife, but Lambrecht (1958) relates from Mayoyao, further north in the province, that these rites also include the shooting of the pigs with a bow and arrow.

The ritual actions of the *mumbā'i* are thus clearly informed by hunting practices, and one might say that the *mumbā'i* enact the spirits as on a hunting expedition and that the *bābuy* given or sacrificed to them by humans are for the spirits the *lāman* they hunt for. The double enactment of wild and domestic that is performed in hunting rituals is thus operative in the sacrifice rituals, as well.[5] If we see sacrifice and hunting as enactments of human-animal-spirit assemblages, they appear as inversions of each other. A human hunt is a spiritual sacrifice. The spirits give up some of their animals, and the hunters reciprocate this by returning the bones of the hunted animals. A human sacrifice is the spiritual hunt. The humans give up some of their animals, and the spirits reciprocate by returning the *lennāwa* they have taken, which in this case was Duntugan's *lennāwa*.[6]

Hence, when we take the cosmological dimension of Ifugao human-animal relations into account, the differentiation between wild and domestic animals gets complicated, as pigs seem to be *bābuy* and *lāman* at one and the same time, although from different perspectives. This double enactment is, however, not constant and changes with context. The *bābuy* that people keep as part of their household only become doubly enacted as spirits' *lāman* through the ritual practices of the *mumbā'i*. Moreover, the partial containment of the *bābuy* entails also that the *bābuy* may move between the human village domain and the spiritual forest domain, and this mobility is crucial for their potential status as *lāman* for the spirits.

A few words on the notions of wild pigs and domestic pigs are warranted here, then. When I have translated *lāman* and *bābuy* as "wild pig" and "domestic pig," respectively, this is partly because this is how my Ifugao friends translated these categories for me.[7] But this should not lead us to think that an Ifugao wild animal is free of control and containment. The wild-domestic distinction here is rather one that operates upon a difference of ownership and perspective. For humans, *lāman* are wild pigs, but the "wilderness" to which these animals belong is the home sphere of the spirits. Strictly speaking, then, all animals are domestic in one way or another. They are all owned by either humans or spirits, although in neither case does that ownership entail complete control over them, and each category may be transformed into the other. And that is a point worth stressing here. Since *bābuy* are hardly ever killed and eaten unless they are sacrificed, *bābuy* husbandry is targeted toward making them available for this transformation. They will eventually all become ritually transformed into *lāman*, and that capacity for transformation requires that they are treated as *bābuy*, that is, allowed to roam around, fed weeds from the fields, and kept close to the house at night.

There is one further point worth mentioning here, and that is that the cosmological dimension of Ifugao human-animal relations also gives rise to an interesting connection between risk and domestication (Clark 2007). Both the improper enactment of human-*lāman* and human-*bābuy* relations may cause illness and death to humans. If successful hunting is not followed up with the return of the bones to the forest, the spirits may take revenge by stealing the *lennāwa* of humans. This may also happen if humans do not maintain the landscape properly, by, for instance, letting the rice terraces deteriorate, and do not treat their pigs according to the "ways of the ancestors."

In addition, the sacrificial ritual in itself is dangerous to humans. When the spirits are brought to the house during the rituals, they may hunt not only for pigs but occasionally also for humans and their *lennāwa*, thus causing the lethal *lennāwa-odol* separation. Hence, the perspectival differences between humans and spirits do not imply that these differentiations are stable and distinct. Indeed, the potential for transformation is always there, as spirits may change their *odol* and appear in both human and animal "clothing," and humans may become transformed into spirits. This potential for transformation means that human is a constant becoming that relies on the proper enactment of relations with spirits and, as part of that, with the landscape and animals.

The practices of keeping *bābuy* the correct way and the double enactment of *bābuy* and *lāman* in hunting and sacrifices thus become part of the effort to stabilize oneself as human, of keeping the potential for transformation at bay (Remme 2017). There is, then, something to what James Scott has claimed, namely, that "in domesticating much of the natural world, we, in turn, domesticated ourself" (2011, 186), but it is also worth adding here that these domestication practices in a way also include a domestication of the spirits, although both of these forms of domestication must be characterized by the same partial containment that pigs enjoy.

From Domestic *Bābuy* to Native Pigs

As I have shown above, *bābuy* and *lāman* and the partial and blurry distinctions between them are central to the enactment of relations between humans and spirits and are implicated in the maintenance of the landscape in which Mahhig, Bugan, Duntugan, and their covillagers dwell. It is only *bābuy* that are bred and fed in a way that makes them usable as sacrificial animals. Hence, both the pig we carried to the wedding and those sacrificed at Duntugan's ritual had to be and were *bābuy*, domestic pigs. Duntugan had not bred and raised the pigs himself but had bought them at the market in the lowland town of Lamut, where *bābuy* are for sale. At this and other markets in the area, another kind of pig is available, although these cannot be used for sacrificial purposes. These are the *bangyādu*, also referred to as "commercial pigs" that some pig farmers breed in large pigpens and that are for sale also at the weekly Saturday market in the nearby town of Banaue. Commercial pigs are of a different breed than *bābuy* and *lāman* and are distinguishable by their pink color and also by being less tasty.[8] Moreover, they are products of an entirely different form of domestication. They are not allowed to roam around freely. They are kept in large concrete pigpens, and they are fed various forms of nutrition-enhanced fodder. The investment costs that this mode of pig husbandry entails inhibits most villagers from engaging in it, but in Banaue and the regional capital, Lagawe, there are people with enough financial resources to have established small commercial pig farms. The commercial pigs do not receive fodder from rice terraces or swiddens, and neither are they kept close to the houses where their owners live. No taboos apply and no *pūdung* knots are tied when piglets are born. However, farmers of commercial pigs do have to follow a set of rules, and the practices that these rules shape have consequences for the way humans and pigs interact and for how the pigs are used in relational enactments.

While the *bābuy* husbandry in the village is governed by the rules the ancestors laid down, the commercial pig industry is regulated by the Bureau of Animal Industry under the Department of Agriculture. The on-the-ground operations of these regulations lie in the hands of the Municipal Agricultural Officer of the Local Government Unit, who certifies and accredits farmers and occasionally monitors the animals' health and controls breeding practices and pedigree records. The regulations require, for instance, that pigpens to be built at a distance from houses, thus enforcing a separation of the commercial pigs from the *baluy* in which people live. The breeding and pedigree control, which is linked to the National Genetic Resource Improvement Program's aim to produce and distribute genetically superior animals, also entails that pigs must be kept strictly within the confines of the cemented pigpens and not allowed to interbreed with the native variants. Accreditation also requires the farmers to keep feeding records, which again requires a feed variant that is easily calculable. Nutrition-improved pig feeds serve this purpose well, much better than occasional weeds and leftovers, and they have the additional commercial advantage of accelerating the growth and fatting of the pigs. Commercial pig farming and the quantification of pigs that it involves thus imply significantly more control of animals than traditional *bābuy* pig husbandry. However, as Anneberg and Vaarst point out (this volume), regulations are not always strictly complied with, as farmers in practice evade legislation and make compromises between care and control.

In any case, these factors make food waste and weed resources from the terraces unsuitable for commercial pigs and thus contribute to separating the commercial pigs from the dwelling practices that make *bābuy* suitable as sacrificial animals. The commercial pigs are never implicated in rice cultivation the way *bābuy* are, and they are exempt from the daily life of the *baluy*. However, this does not mean that they are without significance. Quite the contrary, commercial pigs are in high demand, and their mode of domestication makes them relatively cheap, compared to native pigs. They are therefore used a lot in *madāwat* in marriages as well as in the lavish marriage parties when the entire village is fed pork contributed mainly by the groom's family. During the campaign period for the 2004 municipal elections, the demand for pigs skyrocketed, as all the contenders for the municipal council positions had to provide pigs to attract people to their campaign meetings. In addition, there has always been a close connection between pigs and prestige in Ifugao. In sacrifice rituals, the number of pigs killed always reflects the prestige rank of the family. Prestige is, however, always up for negotiation, and when people in the area are now increasingly integrated in a monetary economy and have access to wage labor, many convert their monetary wealth into prestige by adding commercial pigs to their sacrifices.

The introduction of *bangyādu* has not, I should emphasize, replaced traditional Ifugao pig husbandry. Rather, it takes place side by side with traditional *bābuy* husbandry. This is, then, yet another case that undermines the epochal approach to domestication. Commercial pig husbandry involves greater control and containment, but it has displaced neither human-*bābuy* nor human-spirit-*lāman* relations. In fact, while *bangyādu* production has become an important industry in the area, it has also strengthened local *bābuy* husbandry.[9] The presence of a cheaper alternative has increased the value of *bābuy*. They are considered the only ones considered suitable for sacrifice, but they are also held to taste better, to be more resistant to parasites and diseases, and to be cheaper to raise, in terms of both fodder and housing.

These characteristics of *bābuy* make them particularly attractive for rural smallholders. The Philippine state has recently drawn attention to the potential of small-scale farming of *bābuy* for alleviating poverty, for making the country's pig production environmentally friendly and sustainable, and for counteracting the gradual loss of genetic diversity in the nation's endemic pig stock.[10] In 2010, the Philippine Native Animal Development Program was established, and the Bureau of Animal Industry now heavily promotes the production of *bābuy*, which in this context is referred to as "native pigs." In 2015, the Provincial Governor's Office of Ifugao followed up and funded a native pig farming project. What implications this will have for the forms domestication of *bābuy* in the area is uncertain, but a monitoring regime of native pig husbandry seems to be on track under supervision by the Provincial Veterinary Office, although this supervision must be negotiated with the rights given to the villagers by the Indigenous Peoples Rights Act to conduct pig farming according to their cultural traditions.

The introduction of commercial pigs and the changing conditions of native pig production indicate clearly that forms of domestication change over time and are entangled in processes that extend beyond human-animal relations. The increasing engagement in monetary economy and capitalist ventures combine with various state-driven attempts to control and modify native pig production, and together these factors shape Ifugao forms of domestication. One could be tempted to argue that both commercial and native pig farming seem to become more controlled and monitored than they used to be, thus changing the partial quality of containment that is so crucial to traditional pig husbandry in Ifugao. But the promotion of native pig farming entails also that the flow of fodder from *ūma* and *payaw* can continue and that pigs can be kept close to the *baluy* and become a part of family life. Native pigs under these new domestication forms still qualify as suitable for sacrifices and continue therefore to be of crucial importance in the human-spirit-landscape assemblages.

However, this assemblage in itself is changing. In the last twenty years or so, various Pentecostal congregations have been established in the region and have influenced the enactment of multispecies assemblages, including relations between humans, spirits, and pigs. Those who have converted to Pentecostalism tend to shy away from everything connected with the traditional ritual practices, and the spirits have been redefined as demons. Sacrificial rituals in which people interact with the spirits are therefore a definite no-go zone for the Pentecostals, and receiving and eating meat from sacrificed pigs is prohibited. The most serious implication of this is that the Pentecostals thereby exclude themselves from participation in the exchange of pork that is absolutely central to Ifugao "kinning" (Howell 2003) practices. But conversion also has consequences for landscape transformation. Most clearly is this seen in the Pentecostals' refusal to follow the orders of the agricultural chief, the *tonong*, of the village. Successful rice cultivation in the irrigated terraces requires a careful synchronization of seedling planting, since this minimizes the period in which the fields are vulnerable to pests and insect attacks. This synchronization is secured through the *tonong*, who decides when to start planting. The *tonong* also maintains close contact with the spirits and is supposed to perform yield-enhancing rituals in the village's ritual granary. For Pentecostals, following the *tonong*'s decisions amounts to accepting the spirits' influence, and many Pentecostals therefore plant whenever they want to. The consequence is a lack of synchronization of the rice maturation, which gives birds, insects, rats, and other pests a prolonged feeding period. Rice yields decrease and the rat and pest attacks increase. The former causes farmers to eventually abandon their fields and seek labor elsewhere. The latter does severe damage to terrace walls. Rats and worms dig their tunnels into the walls, and when these tunnels are filled with water during heavy rainfall, the walls eventually burst and lay the fields in ruins.

Conclusion

When we returned to our village after carrying the pig to Wigan's parents-in-law, we followed the path that winds up and down the steep mountainsides. Along the way, we saw the impact of different kinds of domestication practices the Ifugao have engaged in for centuries; swidden gardens where they cultivate a variety of vegetables; irrigated fields where rice is grown along with a host of other greens, fishes, and snails; and forests where they occasionally hunt and keep fruit trees. The coexistence of these different forms of plant cultivation is a sign of the complexities of Ifugao domestication practices and a key to my approach to seeing these practices not as epochal changes but as performed alongside each other.[11] Moreover, all these forms of domestication are in various ways entangled with the enactment of relations between humans and pigs and between humans and spirits, relations that are undergoing change due to socioeconomic changes, state initiatives, and religious conversions.

To the extent that there is an Ifugao domestication story, then, that story is certainly not a linear one but rather one that traces several intersecting lines that continuously move back and forth, twist and turn, and fold and unfold, thus continuously extending and contracting the scope of their more-than-human sociality. These entangled forms of domestication thus decenter the distinctions between wild and domestic and the centrality of control and containment that we tend to mobilize when telling the story about humans and their animals. Domestication in Ifugao is all about partial containment and permeable boundaries, about shifting relations, perspectives, transformations, and coexisting forms of human-animal-plant-spirit relations. It engages human-animal relations in a wider more-than-human sociality that may or may not include spirits and that often entails a risk on the part of humans. Moreover, the cosmological frame in which these human-animal-spirit relations are enacted forces us to rethink what animals are, that is, what ontological status they have and what they are transformed into, for the people involved. Ifugao domestication processes and the variety of shifting and coexisting forms of human-animal-spirit relations thus decenter classical domestication narratives by evidencing alternative stories and relations of domestication and alternative ontological grounds on which human-animal relations operate.

Although the spirit aspect of Ifugao domestication may be less relevant elsewhere, the points presented here can inform our understanding of other domestication processes. The partial quality of control and containment seems to be central in the free-ranging animal husbandry and debates over animal welfare, for instance, and the implication of changing forms of domestication and human-animal relations in wider economic, political, and religious processes is a story worth telling in other contexts, too. The risk element, which is an integral part of Ifugao domestication, is particularly worth noticing here. While classical narratives of domestication emphasize human domination, the risk that domestication involves for humans is often undervalued. The Ifugao case makes it evident, however, that processes of domestication may be both highly productive and dangerous for humans, animals, and whatever other entities are involved in them.

NOTES

1. I refer to Ingold's theorization of domestication here not to support it but to offer it as a reference based on an epochal approach to domestication. Ingold's theory has been critiqued on various grounds, for instance, by Knight (2005) and Oma (2010), who question Ingold's argument about the relations of trust and reciprocity between hunter and prey, particularly since hunting cannot establish any repetitive engagements with the same animal and since there is therefore a significant difference between human-human relations and human-animal relations among hunter-gatherers. In addition, the proximity of humans and domestic animals is dependent on trust and cooperation, which goes directly against the ideas presented by Ingold.

2. Ingold, for instance, emphasizes these features of domestication as he sees it as forms of human engagement with animals based on domination with clear parallels to slavery (2000, 73; Tapper 1988). The emphasis on human control and domination has, however, has been considerably critiqued by approaches that emphasize symbiosis, mutuality, and the active role of animals in human-animal relations (O'Connor 1997; Rindos 1984; Clark 2007; Coppinger and Coppinger 2002), and it has been argued that domestication may also involve a risk on the part of humans (Clark 2007, 63–66), thus problematizing the ostensible hierarchy in the relations between humans and domestic animals.

3. There are also indications that the hunt was connected with courtship (see Willerslev 2007), as some people owned charms used for attracting both wild pigs and women (Barton 1963, 106).

4. To complicate matters more, certain ritual occasions require that the two terms, *bābuy* and *lāman*, are substituted, thus intentionally blurring the distinctions between them.

5. This double enactment is, as I have described in detail elsewhere (Remme 2016), part of a larger series of spatial, temporal, and perspectival indeterminations created in the ritual.

6. It is worth noting here that during the ritual, the *mumbā'i* appeal to the spirits to stop "spearing them" and that deaths are caused by "the spirits hunting us."

7. It is also consistent with Newell's translation in his *Batad Ifugao Dictionary* (1993).

8. *Bābuy* and *lāman* are both variants of Philippine origin (*Sus philippensis*), while the *bangyādu* come from imported breeds of *Sus domesticus*.

9. See Weiss (2016) for a similar kind of relation between industrialized and artisanal pasture-raised pig husbandry in North Carolina.

10. Since the late 1990s, the National Swine and Poultry Research and Development Center (NSPRDC) has run a selection program for maintaining and developing the best

genetic resources for native pig production. Through this program, NSPRDC has developed the so-called BT Blacks, a breed of native pig coming from Benguet, Marinduque, and Quezon provinces, which they aim to distribute to farmers. They also encourage, and assist in the improvement of, breeding other types of native pigs.

11. The existence of swiddens and irrigated rice cultivation is noted also by Acabado (2012, 2010), who sees this form of "composite agricultural system" (Rambo 1996) as a risk minimization strategy. As I have noted elsewhere (Remme 2014b), however, scholars have tended to overlook the centrality of pigs in Ifugao sociality, and I have tried here to emphasize a similar "composite animal domestication system," although without making assumptions regarding risk minimization. Rather, seeing these practices within a cosmological frame, these practices are always accompanied by a certain risk, since they involve interacting with spirits.

REFERENCES

Abramson, Alan, and Martin Holbraad, eds. 2014. *Framing Cosmologies: The Anthropology of Worlds*. Manchester: Manchester University Press.

Acabado, Stephen. 2010. "Landscapes and the Archeology of Ifugao Agricultural Terraces: Establishing Antiquity and Social Organization." *Hukay: Journal for Archeological Research in Asia and the Pacific* 15: 31–61.

Acabado, Stephen. 2012. "The Ifugao Agricultural Landscapes: Agro-cultural Complexes and the Intensification Debate." *Journal of Southeast Asian Studies* 43, no. 3: 500–22.

Barton, Roy Franklin. 1946. "The Religion of the Ifugaos." *American Anthropologist* 48, no. 4, part 2: 1–211.

Barton, Roy Franklin. 1963. *Autobiographies of Three Pagans in the Philippines*. New Hyde Park: University.

Bökönyi, Sandor. 1989. "Definitions of Animal Domestication." In *The Walking Larder: Patterns of Domestication, Pastoralism and Predation*, ed. Juliet Clutton-Brock, 22–27. London: Unwin Hyman.

Cassidy, Rebecca. 2007. "Introduction: Domestication Reconsidered." In *Where the Wild Things Are Now: Domestication Reconsidered*, ed. Rebecca Cassidy and Molly Mullin, 1–26. Oxford: Berg.

Clark, Nigel. 2007. "Animal Interface: The Generosity of Domestication." In *Where the Wild Things Are Now: Domestication Reconsidered*, ed. Rebecca Cassidy and Molly Mullin, 49–70. Oxford: Berg.

Coppinger, Roger, and Lorna Coppinger. 2002. *Dogs: A New Understanding of Canine Origin, Behaviour and Evolution*. New Haven: Yale University Press.

de la Cadena, Marisol. 2010. "Indigenous Cosmopolitics in the Andes: Conceptual Reflections beyond 'Politics.'" *Cultural Anthropology* 25, no. 2: 334–70.

de la Cadena, Marisol. 2014. "The Politics of Modern Politics Meets Ethnographies of Excess through Ontological Openings." *Cultural Anthropology Online*. https://culanth.org/fieldsights/471-the—politics-of-modern-politics-meets-ethnographies-of-excess-through-ontological-openings.

Ducos, Pierre. 1989. "Defining Domestication: A Clarification." In *The Walking Larder: Patterns of Domestication, Pastoralism and Predation*, ed. Juliet Clutton-Brock, 28–30. London: Unwin Hyman.
Fuentes, Agustín. 2007. "Monkey and Human Interconnections: The Wild, the Captive, and the In-between." In *Where the Wild Things Are Now: Domestication Reconsidered*, ed. Rebecca Cassidy and Molly Mullin, 123–46. Oxford: Berg.
Howell, Signe. 1989. *Society and Cosmos: Chewong of Peninsular Malaysia*. Chicago: University of Chicago Press.
Howell, Signe. 2003. "Kinning: The Creation of Life Trajectories in Transnational Adoptive Families." *Journal of the Royal Anthropological Institute* 9, no. 3: 465–84.
Ingold, Tim. 2000. *The Perception of the Environment: Essays on Livelihood, Dwelling and Skill*. London: Routledge.
Knight, John. 2005. "Introduction." In *Animals in Person: Cultural Perspectives on Human-Animal Intimacy*, ed. John Knight, 1–13. Oxford: Berg.
Lambrecht, Francis. 1958. "The Mayawyaw Ritual, No. 7: 'Hunting and Its Ritual.'" *Journal of East Asiatic Studies* 6, no. 1: 1–28.
Law, John, and Marianne Elisabeth Lien. 2013. "Slippery: Field Notes in Empirical Ontology." *Social Studies of Science* 43, no. 3: 363–78.
Lien, Marianne Elisabeth. 2007. "Domestication 'Downunder': Atlantic Salmon Farming in Tasmania." In *Where the Wild Things Are Now: Domestication Reconsidered*, ed. Rebecca Cassidy and Molly Mullin, 205–25. Oxford: Berg.
Lien, Marianne Elisabeth. 2015. *Becoming Salmon: Aquaculture and the Domestication of a Fish*. Berkeley: University of California Press.
Newell, Leonard E. 1993. *Batad Ifugao Dictionary with Ethnographic Notes*. Manila: Linguistic Society of the Philippines.
O'Connor, Terry P. 1997. "Working at Relationships: Another Look at Animal Domestication." *Antiquity* 71, no. 271: 149–56.
Oma, Kristin Armstrong. 2010. "Between Trust and Domination: Social Contracts between Humans and Animals." *World Archeology* 42, no. 2: 175–87.
Povinelli, Elizabeth. 2014. "Geontologies of the Otherwise." *Cultural Anthropology Online*. https://culanth.org/fieldsights/465-geontologies-of-the-otherwise.
Rambo, Terry. 1996. "The Composite Swiddening Agroecosystem of the Tay Ethnic Minority of the Northwestern Mountains of Vietnam." In *Land Degradation and Agricultural Sustainability: Case Studies from Southeast and East Asia*, ed. Aran Patanothai, 43–64. Khon Kaen: Southeast Asian Universities Agroecosystem Network and Khon Kaen University.
Remme, Jon Henrik Ziegler. 2014a. "A Dispositional Account of Causality: From Herbal Insecticides to Theories of Emergence and Becoming." *Anthropological Theory* 14, no. 4: 405–21.
Remme, Jon Henrik Ziegler. 2014b. *Pigs and Persons in the Philippines: Human-Animal Entanglements in Ifugao Rituals*. Lanham: Lexington.
Remme, Jon Henrik Ziegler. 2016. "Chronically Unstable Ontology: Ontological Dynamics, Radical Alterity, and the 'Otherwise Within.'" In *Critical Anthropologi-*

cal Engagements in Human Alterity and Difference, ed. Bjørn Enge Bertelsen and Synnøve Bendixsen, 113–33. Palgrave Macmillan.
Remme, Jon Henrik Ziegler. 2017. "Human at Risk: Becoming Human and the Dynamics of Extended Sociality." In *Human Nature and Social Life: Perspectives on Extended Socialities*, ed. Jon Henrik Ziegler Remme and Kenneth Sillander. Cambridge: Cambridge University Press.
Rindos, David. 1984. *The Origins of Agriculture: An Evolutionary Perspective.* New York: Academic.
Scott, James C. 2011. "Four Domestications: Fire, Plants, Animals, and . . . Us." Accessed February 15, 2013. https://tannerlectures.utah.edu/_documents/a-to-z/s/Scott_11.pdf.
Tapper, Richard. 1988. "Animality, Humanity, Morality, Society." In *What Is an Animal?*, ed. Tim Ingold, 47–62. London: Unwin Hyman.
Tsing, Anna. 2013. "More-Than-Human Sociality: A Call for Critical Description." In *Anthropology and Nature*, ed. Kirsten Hastrup, 27–42. New York: Routledge.
Vitebsky, Piers. 2005. *Reindeer People: Living with Animals and Spirits in Siberia.* London: HarperCollins.
Viveiros de Castro, Eduardo. 1998. "Cosmological Deixis and Amerindian Perspectivism." *Journal of the Royal Anthropological Institute* 4, no. 3: 469–88.
Weiss, Brad. 2016. *Real Pigs: Shifting Values in the Field of Local Pork.* Durham: Duke University Press.
Willerslev, Rane. 2007. *Soul Hunters: Hunting, Animism, and Personhood among the Siberian Yukaghirs.* Berkeley: University of California Press.
Willerslev, Rane, Piers Vitebsky, and Anatoly Alekseyev. 2014. "Sacrifice as the Ideal Hunt: A Cosmological Explanation for the Origin of Reindeer Domestication." *Journal of the Royal Anthropological Institute* 21, no. 1: 1–23.

3

DOG EARS AND TAILS

Different Relational Ways of Being with Canines in Aboriginal Australia and Mongolia · *Natasha Fijn*

Why is it that, in zoological terms, Australian dingoes are classified as "wild," while Mongolian dogs are classified as "domesticated," when both have been intimately associated with humans? This chapter compares two ethnographic cases featuring canine-human relations that are ordinarily classified on opposite sides of the domestic-wild continuum: Yolngu in Aboriginal Australia engage with dingoes that are typically classified as wild; while herding communities in the Khangai Mountains of Mongolia have dogs that are typically viewed as domesticated.

Such distinctions have been made based on physical markers on dog bodies, such as the shape of ears, tails, and muzzles. While there are indeed differences in the bodies of Mongolian dogs and Australian dingoes, this chapter demonstrates how such binary classifications obscure as much as they reveal. Its goal is to show how the metaphysics and everyday perspectives of these two human groups shape the biosociality of canine groups in distinct ways—ways that cannot be grasped through a binary, morphology-based categorization of dogs as either "wild" or "domestic."

To date, much of the work on animal domestication has homogenized both the humans and the communities of animals with whom they are relating, ignoring social and cultural difference. Detailing how human worldviews are intertwined with everyday practices, and in turn, how everyday relations with canines shape dog bodies, I suggest that morphological differences cannot be read simply as markers of different degrees of domestication. Instead, I interpret these bodily signs as evidence of a much more nuanced set of relations, reflecting processes that are both ontological and ontogenetic. This chapter is primarily a comparative analysis of two distinct ways of being with dogs; secondarily, it may be read as a critique of conventional classifications of human-dog relations that tend to prioritize morphological traits as the key markers of dog domestication.

Unlike many parts of the world, where animals have been used to till land, Khalkha Mongolian herders and Yolngu of Arnhem Land have not sought to create settled forms of agriculture. Humans and nonhumans roam freely, rarely bounded by fences; neither group imagines categorical boundaries between the domestic and the wild, or culture and nature. Yet, despite these general similarities, Yolngu and Mongolian have markedly different relations with canines, in part due to differences in ontological worldviews and disparate lifeways.

In the comparison that follows, I draw on personal experience from both sites. In Mongolia, I conducted ethnographic fieldwork while living within a hybrid multispecies community in the Khangai Mountains for a year in 2005, as well as in the spring of 2007.[1] I am also drawing on personal experiences within an Aboriginal homeland in northeast Arnhem Land, primarily during the dry seasons from 2012–2014. For the latter case, however, my approach is also historical. In particular, I draw on the ethnography and natural history accounts of Donald Thomson, whose detailed field notes on the relations between dingoes and humans have rarely been referred to in the existing literature on canine domestication. Thomson studied Yolngu communities in the same field location as myself, but at a time when there was minimal crossbreeding between dingoes and dogs of European origin.[2] This is not because I think that "pure" dingoes are inherently more worthy of study but because I am specifically interested in the influences of Aboriginal ontology and philosophy in relation to dingo biosociality.

For Yolngu, in the homeland communities of northeast Arnhem Land,

Australia, hunting and gathering remain significant to their way of life, and their surroundings include engagement with totemically significant species such as crocodiles, honeybees (sugarbag), turtles, snakes, and sharks (see Fijn 2013, 2014). In Mongolia, the herders who live in the Khangai Mountains remain seminomadic, while their lives are fundamentally intertwined with horses, cattle (including yak), sheep, and goats (see Fijn 2011). Metaphysical schemes and philosophy matter in the daily practices of living with such animals, including the canines, who are significant entities within these two very different communities.

How do the ontologies of these communities, as enacted within everyday practice, come to shape canine ears and tails in different ways? (figure 3.1). Human-canine relations involve different processes of development in both an ontological sense and an ontogenetic sense. While physical traits, such as the ears, tail, and muzzle, are often read as distinguishing characteristics of domestic or wild dogs, my intention is to show that these bodily signs are evidence of much more nuanced differences in human-dog relations that cannot be described by segmented binaries.

Domestication and Dogs' Bodies

In zoology and bioanthropology, tameness is not the same as being domesticated. A tame but nondomesticated individual displays many resemblances with animals in the wild. Upon reaching adulthood, it typically disperses to breed rather than reproducing as an isolated population within a domestic environment. Thus, tameness does not necessarily involve control over reproductive behavior.[3]

Domestication, on the contrary, is defined by its specific effects on animal bodies, which are assumed to be a product of reproductive control (see introduction). Zoologists Wilkins and colleagues (2014) have identified a suite of heritable traits in domesticated mammals and have labeled this suite of characteristics a "domestication syndrome." Characteristics associated with this "domestication syndrome" include behavioral tameness and docility, changes in external appearance (such as depigmentation in the form of white coat color and markings), decreased tooth size, changes to the skull, floppy or reduced ears, and curly tails. Changes in hormonal levels are also evident, along with a retention of juvenile characteristics, including vocalizations toward humans (i.e., barking, in the case of dogs).

As these zoologists point out, the idea that domestication creates changes in bodies and temperament is not new. Charles Darwin (1868) recognized that

FIGURE 3.1. Mongolian dog with characteristic floppy ears, yellow eye markings, and predominantly black coat coloration. Photo by Natasha Fijn.

domesticated animals exhibit particular traits, but he did not have an answer as to how or why such traits occur. Wilkins and colleagues (2014) drew upon a famous Russian experiment from the late 1950s on the domestication of silver foxes, the work of a scientific team led by geneticist Dmitry Balyaev. This classic fox farm experiment found that, when selected specifically for tameness and docility, foxes also exhibited changes in physiological traits, such as floppy ears and a curled tail (Trut 1999). Balyaev's hypothesis was that when animals are selected primarily for behavioral traits, physiological, hormonal, and morphological change is evident as a secondary result (Belyaev 1979). What is new in Wilkins et al.'s proposal is their location of the source of these traits in changes in the neural crest cells during embryonic development. In zoological terms, these changes occur as part of the initial stages of development of an individual, as a part of ontogeny (i.e., part of the development of the individual organism).

These kinds of changes are consistent across mammal species but are particularly evident in canines. Mongolian dogs have floppy ears, a curled tail, a shortened muzzle, and white patches on their fur—all attributes aligning with domestication syndrome characteristics, and therefore, they would be classi-

FIGURE 3.2. Unidentified Yolngu boy holding two dingo puppies, Eastern Arnhem Land, Northern Territory, 1942. Note the dingoes' characteristic erect ears, long muzzle, consistent coat color, and lack of white patches. Photo from the Donald Thomson Ethnohistory Collection. Credit: The Thomson family and Museum Victoria.

fied by zoologists as domestic (figure 3.1). However, as I will detail below, Mongolian dogs are *not* particularly selected for tameness and docility, as their role is to defensively guard the encampment. Hence, the selection criteria are different from those that were applied within the Russian experiment.

Based on similar classification practices, zoologists typically categorize dingoes as wild. In terms of Linnaean taxonomy, the dingo is recognized as a separate subspecies, *Canus lupus dingo,* in contrast to the familiar common dog, *Canus lupus familiaris.* Indeed, dingoes do not seem to exhibit many of the bodily attributes of domestication syndrome. In accordance with the different environmental conditions on the Australian continent, dingoes have larger canine teeth, their muzzles are longer, their tails are not curled, and their ears are large and stand upright (figure 3.2). They breed only once a year and often forage and hunt in packs. They do not often bark but instead howl like wolves and do not whimper as much as other dogs when hurt (Thomson 1957).

In experiments on dingoes' responses to human social cues, such as those in which humans point to a hidden food reward, dingoes behave in ways conventionally associated with less domesticated species. In one experiment, a few dingoes would not remain in the enclosed testing scenario and were too afraid to be tested at all. Those that were more socialized toward humans from a young age (and were still under twelve months of age) responded more readily to humans than wolves did but were not as attuned to human cues as domesticated dogs were (Smith and Litchfield 2009). Based on such experiments, dingoes are classified as *behaviorally* dissimilar from wolves, yet not as dependent upon humans as European-derived domestic dogs.

In what follows, I shall argue that the classification of dingoes and Mongolian dogs into the categories of domestic-wild misses the importance of human cultural differences in the making of dog bodies. More precisely, these binary distinctions mask the culturally specific practices that give rise to such different physical traits and ignore the human-dog relations with which they are intertwined. As will be outlined below, canines and humans can have vitally significant relations *without* intensive reproductive selection and control, or selection for tameness and docility. Such relations are intertwined with the development of particular morphological traits, but not, as I will show, in the ways many scholars of domestication imagine. My argument, in short, is that a less Eurocentric paradigm—one that does not begin from Western notions of domestication as reproductive control and artificial selection—can help us to better notice and understand differences in canine bodies, as well as the diversity in human-dog relations.

Yolngu-Dingo Relations

To address this point, I want to begin by describing social relations among Yolngu and dingo, in historical perspective. To do so, I begin with a note about the definition of the dingo. Prior to the arrival of European fleets in 1788, the dogs residing in Aboriginal communities were a distinct canine kind, what is now referred to as the dingo. Colonization brought not only European settlers but also the European domestic dog, which Aboriginal peoples quickly assimilated into their camps. The term "dingo" was first adopted into the English language around the time of settlement from the name for tame camp dogs within an Aboriginal community in New South Wales (Ryan 1964). Dingo is now often used interchangeably with "wild dog" among non-Aboriginal Australians, who generally perceive dingoes as illusive wild creatures that often survive by killing livestock. Among farmers, the dingo has especially negative connotations, in contrast to the domestic Australian sheepdog, who is seen as a partner or mate. Among Australian Aboriginals, however, these words take on different meanings. In Aboriginal Dreaming ceremony and narrative, the words “dingo” and “dog” are used interchangeably, as all canines are derived from dingo/dog ancestor dreaming (see Parker 2006). There are many words for dingo/dog in Yolngu-matha, some of which are sacred and cannot be spoken, some describing those that live in the bush, and some for those that live within a community.

The question of who is a dingo is clearly far from simple. The canines found in homeland communities today consist of a variety of different breeds, often retaining some of the distinctive features and characteristics of the dingo. These mixed breeds are not isolated populations and have been influenced by selective breeding, previously bred in accordance with the domestication and colonial perspectives of broader Australian society. For the purposes of this chapter, however, I am using the term “dingo” to refer to the canine that Aborigines lived among prior to European settlement, rather than to Yolngu relations with the current mix of different dog breeds residing within Aboriginal communities today.

WHEN THE ANTHROPOLOGIST and naturalist Donald Thomson was in the field in Arnhem Land in the early 1930s, he described how an Aboriginal hunter returned with an armful of dingo puppies, captured from a den in the bush. As a caption to the photograph of this hunter, he remarks: “The dingo, unlike the white man’s dog, is strongly nocturnal. It never barks but gives voice to a long melancholy howl like the call of the wolf ancestor in the northern hemisphere. The dingo never becomes truly domesticated and at best gives the Aborigine only a half-hearted allegiance, helping him in his quest for game, and after two or three years in camp, it usually runs wild to find a mate and to breed” (Thomson 1972, 52). Until they reached breeding age, young dingoes remained within the *wanga* (or camp), scavenging from humans, while socializing with children and other young dingoes (figure 3.3). Dingoes were kissed, patted, and coddled as puppies but largely ignored or physically threatened if they came too close as adults. The desire to nurture was lost once the dingoes became adults, and they “quickly progressed from babies to beasts which everyone loathes” (Kolig 1978, 93). When they reached breeding age, dingoes were inclined to return to the bush (see Meggitt 1965; Hamilton 1972; Kolig 1978). Deborah Rose (1992) relates this developmental phase of leaving the camp to the process that Aboriginal males went through when they traditionally left camp upon reaching sexual maturity to participate in male ceremonial activities. As Smith and Litchfield note: “That dingoes were encouraged to leave

FIGURE 3.3. Rraywalla, a Mildjingi man, with dingo pup, Central Arnhem Land, Northern Territory, 1937. Photo from the Donald Thomson Ethnohistory Collection. Credit: The Thomson family and Museum Victoria.

upon sexual maturity . . . goes somewhat against the principle of domestication, which encourages animals to become closely bonded with humans" (2009, 115).

Donald Thomson wrote in his field notes from Arnhem Land in 1932 that two different litters of dingoes he personally raised were not as tame or as docile as the dogs he bred at home: they were more inclined to wild-type fight or flight behavior. At Blue Mud Bay (a bay just south of my field location at Garrthalala), Yolngu hunters brought three dingo puppies from a den. Thomson exchanged sticks of tobacco for the puppies: "They climb much more than domestic puppies, and they are much more savage; they snap quickly, cower at first when annoyed but turn and snap if the annoyance persists, or if they are cornered. They engage in sham fights together but often develop into real fights; they snarl and fight more savagely than white-man dog puppies at the same stage" (Thomson's original field notes, from Dixon and Huxley 1985, 168). I, too, observed this alternate fight-flight behavior in a young part-dingo at Garrthalala. She was initially fearful of my presence in the community and responded by rushing at me, trying to nip at my heels and barking before dashing away. She also responded to a strange dog in the community that was traveling in the back of a vehicle by cowering away from the vehicle and hiding. I initially thought this was unusual behavior for a dog but then realized that this is typical for an animal that has *not* been selected for tameness and docility (see the video segment with this part-dingo: https://vimeo.com/109279575).

In Thomson's field notes, he wrote that one of the pups in the dingo litters he was rearing was growing up to be wild and independent and was causing problems by keeping him awake at night. In a curt manner, he notes how he ended up shooting this dominant one. In the logic of selection for domestication, animals that are too strong-willed and uncontrollable do not survive. Aboriginal Australians have other ideas about rearing relationships and selective killings. Annette Hamilton, who studied Yolngu in the early 1970s, writes that when a Jankutjara man complained of the depredations of his nine dogs, a non-Aboriginal person suggested to him that he shoot some of them. "He turned in horror. 'Kill the dogs?' he exclaimed. 'Do men kill their children when their own hands brought them up?'" (Hamilton 1972, 287). This reference to bringing up dogs in a similar way to children is significant. He is not saying that dogs should be treated *as* humans but that dogs are nurtured in a similar way, resulting in the establishment of close bonds between human and dog. But rather than loyalty and control, this bond incorporates a perspective of dog autonomy and independence as part of the journey toward adulthood.

One of Thomson's accounts illustrates the contrasts between the Yolngu attitude toward young dingo pups and Thomson's own more Eurocentric approach: "Once, when I attempted to pick up one of the puppies, it turned and snapped savagely, and I committed the unpardonable sin of cuffing it across the head. At once I was surrounded by advisors who told me that my methods were quite wrong—that I should have *bitten* the ear of the puppy instead. 'Pay back,' they said—that is, the approved form of correction was to pay it [back] in its own coin" (Thomson 1957, 17). In later field notes, Thomson adds that when a pup growls, Yolngu let it bite their hands and then in return bit the puppy's ear. Instead of trying to dominate the puppy with physical force, the Yolngu attitude is to engage with the puppy as another older dog would, correcting it on canine terms rather than anthropocentric terms.

As Thomson's account demonstrates, dingoes were clearly tamed as puppies yet developed an independence of humans as adults and were subsequently allowed freedom to make their own choices. Yolngu did not attempt to retain dingoes as separate populations, forming isolated genetic lineages. Instead, dingoes were given the freedom to mate with whomever they chose. Tamer, less aggressive individuals could continue to benefit into adulthood through scavenging from scraps and refuse around human camps, while those more fearful of humans could choose to seek sustenance elsewhere (see Coppinger and Coppinger 2001 for the "self-domestication" hypothesis).

Annette Hamilton (1972) similarly emphasizes the parallels between the ways children and dogs are brought up in Yolngu communities: the indulgence of puppies and children; little attempt to discipline and train; a special call for pups and infants; the rubbing of both babies and pups with ochre to protect them from negative spirits; and a delay in the naming of individuals. There is no doubt that there was a deep emotional connection with dingoes, but this was primarily expressed in the nurturing and taming of puppies and not nearly as much with adult canines.

Taming early in life offered behavioral options for dingoes by giving them exposure to human communities, but they were then given the freedom to form their own way of life. Because dingoes were given the freedom to roam, they could establish their own territories and were not solely dependent upon humans for their food supply. Besides scavenging for food, they hunted opportunistically, accompanying Yolngu in their quest for game or within their own hunting packs. Dingoes could and did move into a variety of spaces, including those occupied by humans, changing their behavioral traits in the process. Dingoes demonstrate how canines who relate with humans are not necessarily locked into a trajectory of domestication, understood as human control. Furthermore, they show how reproductive selection was not based on human preferences for individuals that expressed tameness and docility but instead was based upon the Yolngu philosophy of allowing a dog its own agency to choose a mate beyond the borders of the camp.

A more recent example from my own fieldwork further illustrates the Yolngu attitude toward canine autonomy and individual agency. While having a cup of tea in the community of Garrthalala, a Yolngu elder related to me a story about a part dingo that happened to be sitting beside her at the time. As a youngster, the dog had chased after a vehicle when another family was leaving a neighboring homeland community. Somehow the dog navigated the rough road to where the elder lives, even though it was many kilometers away. Instead of rejecting the dog or taking the dog back to the family who had brought her up, the elder kept her within the Garrthalala homeland community. She was given the right to choose where she wanted to stay (see video segment "Garrthalala Dogs" at: https://vimeo.com/109279575).

While dingoes engaged in *significant* relations with people, those relations were cosmologically and materially different from those of European domesticated dogs. Within Aboriginal ecological philosophy, it was not and is not considered appropriate to retain and breed dingoes in a fenced, captive environment. Yolngu give canines (both dogs and dingoes) the freedom to socialize together, as one significant totemic being among a whole raft of Significant Others.[4] Yolngu think of dogs as being part of what could be referred to as a "social ecology," where everything and everyone is interlinked through an extended form of kinship and grounded by the land, or place.[5] As has been noted among other hunting societies (Ingold 2000; Nadasdy 2007), Yolngu generally think of animals on a species level rather than in terms of engagement with specific sentient individuals or individual selves.

To elaborate, a Yolngu elder may speak of a dingo that she encountered in the bush as "My Mother" to other family members nearby. This does not mean that she thinks that the individual dingo is literally her mother, or a reincarnation of her deceased mother (as perhaps a Mongolian herder might). Instead, she is acknowledging that the dingo ancestor is part of her mother's clan's Dreaming. She is also acknowledging the tracks of the ancestral dingo, which are evident on her mother's clan land, as well as the fundamental link between her mother, the dingo, and the land. In other words, her form of social ecology is intricately interwoven in a web of connectedness with land at the center. The totemic connection, therefore, relates to ecological knowledge of a species within a particular area of land, yet at the same time, the dingo ancestor is multiply significant in a symbolic sense.[6]

Ancestral beings are shape-changers: they have the ability to morph into human or dingo. In storytelling, it may be difficult to distinguish whether the protagonist is in the form of a human or in the form of a dingo, and for the purposes of the narrative it does not matter. Deborah Rose, in her book *Wild Dog Dreaming*, refers to a story about the dingo told by Old Tim Yilngayari. He universalizes the story when telling it: "They walk, they stand up, they are finished being dogs now, they're proper humans, women and men. Mother and father dingo made Aboriginal people. White children come out of white dog" (2011, 7). For those individuals with the dingo as one of their totemic species, like Old Tim, he has "an intimate concern for, and understanding of, dingoes and dogs. His life is connected to dingo life and he has assumed a special responsibility for canines" (Rose 1992, 29).

Donald Thomson (1949, 1957) described a ceremony from northeast Arnhem Land associated with the seasonal departure of visiting Macassan fleets from the region of Sulawesi on the Indonesian archipelago. The arrival of the ancestral dingo in association with the Macassan seafarers aligns with current theories of the first arrival of canines onto the Australian continent (some 3,000–5,000 years ago, according to Fillios and Taçon 2016). The ceremony depicts the spirit shadow of the Macassan anchors and incorporates totemic effigies in the form of a dog. In a Yirritja moiety song from Arnhem Land, Fiona Magowan describes the ancestral Yirritja Dog ceremony: "As the Warramiri clan moves from place to place, the hunters throw out the anchor to moor at different locations. Warramiri sing of the anchor as a Yirritja rock called Dhukurrurru, implying the Dog's spiritual presence embedded in the land at each place where the boat is moored" (2007, 140). Magowan explains these totemic interconnections between humans and nonhumans as dwelling within one another. "Yolngu experiences of the environment are shaped by the logic of polymorphic transposition that enables specific attributes of living and non-living things to be interconnected creating a complete system in which they perceive all things to be ancestrally related" (141).

Standard notions of dog domestication do not align with Yolngu social ecology and connection to country, whereby everything and everyone is interconnected through an extended kinship network. This is a cosmological difference that has clearly shaped the selective pressures on canines in these communities, with behavioral and bodily consequences. Without active selection for tame and docile characteristics, dingoes did not develop the suite of traits that are associated with "domestication syndrome." They retained their upright ears, long snouts, and coats without white patches, along with their alertness and wariness. Therefore, the dingoes' lack of domestication syndrome traits does not indicate that dingoes are *not* part of close relations with humans during significant parts of their lives. Rather it is an indication that, due to Yolngu cosmological and ecological philosophy, human-dingo relations required that dingoes retained independent lives as adults, in which they maintained control of their reproductive lives, even within the context of living with humans.

Dogs in Mongolia

Human-dog relations in Mongolia also are dissimilar from those in European worlds, but in different ways from those in Aboriginal Australia. In Mongolia, there are two broad types of dog that predominate: lighter tan-brown hunting dogs (*anch nokhoi*), associated with the forested taiga, and a thickset, *Mongol bankhar,* with a thick, mainly black coat that takes the role of guarding the encampment (*khotch nokhoi*) (although both types still guard encampments). Their stocky physique and thick, woolly coats allow these dogs to survive in the extreme continental climate of Mongolia. The dog's role does not include rounding up herd animals. Herder and horse accompany a combined sheep and goat herd, while dogs remain within the domestic sphere to guard the encampment from intruders.

While Mongolian dogs are actively selected for traits and external coat color patterns, they are not "domesticated" in the European sense, in terms of being artificially selected for tameness and docility. Instead, there is a loose, often unintentional, form of selection for independence and loyalty to the encampment. In Mongolian herding communities, selection for tameness is a common part of the management of animals such as horses, cattle, sheep, and goats. Dogs, however, are treated differently and are subject to a slightly different set of selective pressures. The behavioral preference is for skills in hunting or guarding the encampment. There is a preference for dogs who are capable of being independent with an ability to fend for themselves, resulting in selection through herder selection and through a dog's own choice of mate. Herders value dogs that are *loyal* to one family, guarding against intruders, and they are particularly disparaging about "lazy" dogs not performing their key role in defense of the encampment. This is also the case for members of a herding family, as a real character flaw is if someone is perceived as being lazy and not pulling their weight.

Dogs may occasionally have the opportunity to mate with other dogs from nearby neighboring encampments, but herders make desirable matches between a male and a female. When choosing a dog, a herder will look at a dog's physical appearance, such as coat color. This is seen as a measure of a dog's positive characteristics, for instance, a dog with a heart-shaped white marking on the chest is considered more likely to be loyal. It is interesting that through

FIGURE 3.4. Two Mongolian Bankhar puppies sitting outside a Mongolian *ger* (yurt). They are not allowed to enter the home. Photo by Natasha Fijn.

herding knowledge, a link has been made between an external characteristic, in this instance white pigmentation, and a behavioral trait, such as loyalty, which are the same characteristics that have been bundled together genetically, in accordance with the domestication syndrome hypothesis (Wilkins et al. 2014). In the Mongolian case, however, the external coat color is not selected as a marker for tameness and docility but rather as a marker for loyalty and independence.[7]

Mongolia is popularly known as "the land without fences," and adult herd animals are not restricted in where they can roam when not needed for specific tasks, like riding or milking. Dogs are generally tied up on the outskirts of the encampment and curl up to sleep during much of the day. At night, they wander freely around the encampment to guard the domestic sphere against wolves and rustlers, who may steal precious herd animals. Any scraps or leftovers are placed into a bucket and fed to the dogs, resulting in the dogs subsisting on the same diet as the herding family. Although the dogs are not kept within fences, they do not roam far beyond the territorial boundaries of the encampment (figure 3.4). While young sheep and goats are often nurtured within the family home (*ger*), in my experience, dogs tend to be left to fend for themselves and are not permitted to enter the home. Dogs, then, are engaged with quite differently from the five-kinds-of-animal (*tavan khoshuu mal*) crucial for pastoralism in Mongolia (horses, sheep, goats, cattle, and camels).

In Mongolia, dogs and people are often seen as sharing the same substance but as having the ability through reincarnation to metamorphose into different physical bodies. One ritual illustrates this: When a dog dies, the tail may be cut off to assist the dog to reincarnate into a human in its next life, as humans are one of the few animals without a tail. A Mongolian Buddhist monk told me that one would strive for Nirvana, but if this were unattainable then one would strive to be a human, followed by a dog in the next life. It pays, then, not to treat an individual dog badly, as the dog could be the reincarnation of a beloved family member.

The attribution of individuality or selfhood to dogs is further expressed in the ways dogs are spoken to. An elderly Mongolian herder told me in wonder: "I heard that in countries like yours people talk to dogs as if they are human." It was evident that he thought that this was strange behavior indeed. In a Mongolian herding encampment, dogs are spoken to using specific kinds of vocalizations peculiar to dogs.[8] A dog is not spoken to in complicated sentences but with simple phrases and body language, like the approach used with a preverbal infant. Some words used for Mongolian dogs are also used when addressing Mongolian babies.[9] Speaking to dogs using a specific kind of vocabulary does not deny that the canine has a mind but recognizes that they have a different kind of sociality and behavioral sensibility—one that does not include complex human language.

Unique, independent, and innovative behaviors are valued as important qualities in both dogs and children in Mongolia; they are a key part of selfhood. This requirement for independence is apparent in the naming practices of dogs. As guardians, dogs have a different role from herd animals and are, therefore, named in a different manner. Traditional Mongolian names for dogs are often based on animals that are thought of as having particular strength of character, such as a lion, tiger, dragon, or eagle (see also Bamana 2014). In one encampment, for instance, one dog was named Shonkhor, meaning falcon, while the other was named Ancho, meaning hunter. The names for dogs are often derived from species of animal that feature in folk tales and classic narratives. They are also generally top predators—those animals that are viewed as containing the strength and power that herders would like their dogs to have as protectors of the family.

During fieldwork in Mongolia, I was told that the dog Ancho (the Hunter) was given his name because of his interest and skill in that domain. Of his own volition, he regularly hunted for marmots whenever he spotted one emerging from a hole nearby. Whether dogs are taken out on hunting expeditions in search of deer or boar in the nearby forest depends upon the initiative of the individual dog, rather than upon rigorous training on the part of the herder.

Such practices further demonstrate the ways that dogs are seen as independent and agential beings with their own inherent traits.

As loyalty is highly valued in the Mongolian dog, a dog that deserts the family is considered to have a character fault (also see Irvine 2012).[10] While living with a herding family in the Khangai, one of the dogs did not migrate with us when the family and the herd animals moved from the spring to the summer encampment. I asked what had happened to Shonkhor the dog, but the family seemed unconcerned about his disappearance, and he eventually turned up a few days later of his own accord. This attitude is specific to dogs' disappearance; herd animals' disappearance was a cause for greater concern. When four calves became mixed up with another herd on the same journey, one of the women went out in search of them and guided them to the correct location. As an adult dog, however, Shonkhor was expected to look after himself. This indicates that dogs are seen as beings who are capable of proficiently navigating their worlds without the need for overt human guidance and management.

Conclusion

What is common among Mongolian and Yolngu ways of relating to canines is a sense of inherent *dog competency* and *agency.* Dependency, asymmetry, and control—notions at the heart of Euro-American ideas of domestication—make little sense in relation to canines in either location, where adults are seen as competent and independent beings. Within both Mongolian and Yolngu worlds, puppies are nurtured by humans in much the same way as human infants and are similarly expected to achieve independent, agential adulthood.[11] Yet despite these similarities in the ways in which dogs are granted autonomy and independence, there are nonetheless marked differences between the cosmological orientations of Yolngu and Khalkha that are intertwined with distinctly different quotidian practices.[12] Mongolian herders focus on dogs as individuals with distinctive characteristics and personality, while Yolngu focus on the consubstantiality and extended kinship between themselves and other beings, such as dogs/dingoes, and the ancestral being from which they are both derived.

When comparing Mongolian and Yolngu attitudes toward canines, it is evident that these are different ways of being *with* dogs that do indeed shape dog bodies. Mongolian dogs exist in what I have previously described as codomestic and reciprocal relationships (Fijn 2011), in which humans, dogs, and different species of herd animal dwell together within the same domestic sphere,

FIGURE 3.5. Mongolian dog as part of the codomestic sphere. Photo by the author.

each taking on their own important roles and providing for each other in a reciprocal manner (figure 3.5).

Mongolian herders do not treat their dogs as humans but recognize that dogs have their own social structure and way of being in the world, as individual selves. Dogs are thus given a large degree of freedom to make their own choices and to behave independently, without fenced boundaries. Mongolian dogs are partly assisted in reproductive selection and exhibit physical traits associated with the domestication syndrome, but this cannot be attributed to selective pressures toward tame and docile individuals, as in the Russian fox farm experiment. The dogs may have white marks, but these are seen—and even selected for—as marks of loyalty, not of tameness.

In the Yolngu case, juvenile camp dogs and dingoes are treated with affection and care, often in ways that resemble domestication-type relations elsewhere. But as these canines get older, they are encouraged to be more autonomous, to socialize with their own, and hence to mate freely. Part of the reason Yolngu communities did not retain dingoes to breed in captivity is that it did not, and still does not, run in accordance with Aboriginal ecological philosophy. Dingoes show few (if any) of the classic bodily marks of domestication, but to interpret this as a sign of "wildness" would be to ignore how dingoes coexist with humans. Yolngu cosmology locates the dingo as part of an extended kinship network, or "social ecology." Like Mongolian herders' dogs, dingoes and Aboriginal camp dogs are completely entangled in human worlds, but the ways of understanding and living with canines differs, and hence, the physicality of the canines differs, too.

From these examples, we see how many ways of living together with other species do not necessarily fit into the existing narrow definitions of domestication. A superficial comparison of canines in Mongolia and Arnhem Land may reveal standard physical markers of domestication (domestication syndrome traits) in the former but not in the latter. However, to assume that an individual dog in a Mongolian herding encampment is therefore "more domesticated" (or less wild) than in a Yolngu community is to mask the complexities of what is going on in practice, as well as in people's conceptual schemas. The domesticated/wild binary is clearly an inadequate classificatory tool for thinking about canines cross-culturally, because the options for human-animal relations are not either total control over breeding or a complete absence of human engagement, as European notions have often framed domestication.[13]

These two different ways of knowing dogs in Australia and Mongolia influence the perspectives of individuals living within these two different communities, as well as the everyday relations between their human and canine residents. Different ontological perspectives toward other beings ultimately reflect upon how dogs and humans engage and socialize with one another, which, in turn, inevitably influences ontogeny and a dog's learning and development within its lifetime. In the long term, these differences in cosmology and daily practices also exert selection pressures and ultimately influence the physical bodies of dogs. The result is the morphological variation of dog bodies, as well as biosocial diversity, but to locate this variation neatly within domesticated/wild binaries masks the variation involved in human-canine relations, and the significance of dogs in peoples' lives. Without paying attention to cultural perspectives in relation to domestication processes, we fail to appreciate the nuanced variation in the different ways we coexist with canines.

NOTES

1. See Fijn (2011) for a detailed description of coexistence within a domestic sphere between herders and herd animals in the Khangai Mountains of Mongolia.

2. I am not making a statement against the hybridization of European dog breeds with the dingo (see Probyn-Rapsey 2015) but rather am focusing predominantly on

dingoes during the Thomson era for the simple reason that it is difficult to separate the Aboriginal influence on dingo ears and tails without the overriding influence of Eurocentric domestication, entwined with settler colonialism. The situation is clearer when drawn from Donald Thomson's field notes on dingoes from the 1930s.

3. Elephants are, for example, not considered domesticated, as although they may develop a close working relationship with humans, they are caught in the forest when young and typically do not reproduce within human communities (Clutton-Brock 2012).

4. As biologist Laurie Corbett (1995) wrote, "Wild dingoes, therefore, can be tamed but not domesticated. Should humans [i.e., settler Australians] determine and selectively breed certain standards and characteristics for dingoes, they will cease to be dingoes. A domesticated 'dingo' is not a dingo but just another breed of dog" (cited in Coppinger and Coppinger 2001, 67).

5. According to Paul Nadasdy, neofunctionalist cultural ecologists "eschewed the concept of 'society' in relation to animals altogether in favor of 'population' and 'ecosystem'; relations between people and animals were important, but they were most assuredly not 'social'" (2007, 39n8). Using "social ecology" as a term brings connections between humans and other beings back into the ecological picture.

6. For those that do not have the dingo as their totem, a dingo does not hold the same significance and is just another being in the world, whereas those that do have the dingo as their totem see the significance everywhere, particularly in relation to a mutual connection to country.

7. Herders also actively select for other physical features, such as "four eyes'" (*durun nudtei*), light-brown spots above the eyes that are thought to fend off wolves in the dark, as well as negative spirit forces.

8. I have previously written about vocalizations in relation to herd animals in Mongolia and what I refer to as "multispecies enculturation" (Fijn 2011, 104–28).

9. The only instances that I know of where animals are spoken to in a more complex manner is for ceremonial and ritual purposes, such as in Buddhist chants, using phrases such as the mantra "*Om mani padme hum*." Eduardo Kohn (2007) refers to this kind of altered communication with dogs as a "transspecies pidgin," with its reduced grammatical structure, including both human and dog talk. Among the Amazonian Runa, to "advise" a dog, or to teach it a lesson, the dog's mental state must be altered by administering a hallucinogenic *tsita* mixture. The dog can then be spoken to in prose. The dog essentially becomes a shaman, its interiority changing from dog to human while in an altered state. The Mongolian dog, therefore, can understand the meaning of a ritual through the act of chanting.

10. See the film "Cave of the Yellow Dog" for an instance of this, where the adult herders are portrayed as being reticent to adopt a stray dog and then attempt to leave the dog behind to fend for itself (Byambasuren 2005).

11. This echoes Tim Ingold's ideas that animals are not made but grown: "Bringing up children or raising livestock . . . animals or people are not so much made as grown, and in which surrounding human beings play a greater or lesser part in establishing the conditions of nurture" (2000, 87). In this respect, both the Yolngu and Mongolian cases are

also somewhat similar to human-dog relations found among other nonagricultural peoples. Myrdene Anderson (1986) notes that, like Saami children, reindeer herding dogs learn by doing, not through training: "The reindeer-herding dog, like the Saami child, is treated roughly but affectionately, and given little negative feedback. Instead, being rewarded for unique, independent and innovative behaviour, both dog and child ripen into maturity at their own pace, picking up skills as needed or desired" (10).

12. In a cosmological sense, Mongolian herder perceptions of other beings could be categorized as a form of animism or perhaps analogism, while Aboriginal Australian perceptions could be considered as the ultimate exemplar of totemism. Yet Philippe Descola's (2013) definition of animism requires that the attribution of interiority to nonhumans must be *identical* to one's own. This is quite a specific form of animism and differs from my broader definition of animism in relation to Mongolia (see Fijn 2011, in alignment with Graham Harvey's [2006] definition). What is valuable in relation to Descola's (2013) theoretical approach is the consideration of different human perspectives in relation to the interiority or physicality of other beings, in this instance, Yolngu and Mongolian perceptions about the minds and bodies of dogs.

13. It is also worth considering how this kind of approach may similarly be used to highlight the diverse ways of relating to dogs represented by current Euro-American breeds. Large working dogs (e.g., sled dogs, rescue dogs, and livestock guardian dogs) are quite different, morphologically and behaviorally, from pet dogs, such as Chihuahuas. It is also worth noting that, in contrast to artificial selection practices through human control, in Arnhem Land and Mongolia (and perhaps also in the case of some working dogs, such as sled dogs) adaptation through the selective influence of climatic and environmental factors may outweigh the adaptive pressure from human selection and decision-making. It would be pointless to try to apportion to what degree the dogs are influenced by nature or culture, by human decision-making or environmental factors.

REFERENCES

Anderson, M. 1986. "From Predator to Pet: Social Relationships of the Saami Reindeer-herding Dog." *Central Issues in Anthropology* 6, no. 2: 3–11.

Bamana, G. 2014. "Dogs and Herders: Mythical Kinship, Spiritual Analogy, and Sociality in Rural Mongolia." *Sino-Platonic Papers* 245: 1–18.

Belyaev, D. K. 1979. "Destabilizing Selection as a Factor of Domestication." *Journal of Hereditary* 70: 301–8.

Byambasuren, D. 2005 (film). *The Cave of the Yellow Dog.* Tartan Films, Germany, Mongolia.

Clutton-Brock, J. 2012. *Animals as Domesticates: A World View through History.* East Lansing: Michigan State University Press.

Coppinger, R., and L. Coppinger. 2001. *Dogs: A Startling New Understanding in Canine Origin, Behaviour and Evolution.* New York: Scribner.

Darwin, C. 1868. *The Variation in Animals and Plants under Domestication.* London: John Murray.

Descola, P. 2013. *Beyond Nature and Culture.* Chicago: University of Chicago Press.

Dixon, J. M., and L. Huxley. 1985. *Donald Thomson's Mammals and Fishes of Northern Australia*. Melbourne: Nelson.
Fijn, N. 2011. *Living with Herds: Human-Animal Coexistence in Mongolia*. Cambridge: Cambridge University Press.
Fijn, N. 2013. "Living with Crocodiles: An Alternative Perspective on Engagement with a Reptilian Being." *Animal Studies Journal* 2, no. 2: 1–23.
Fijn, N. 2014. "Sugarbag Dreaming: The Significance of Bees to Yolngu in Northeast Arnhem Land." *Humanimalia* 6, no. 1: 41–61.
Fillios, M. A., and P. S. C. Taçon. 2016. "Who Let the Dogs In? A Review of the Recent Genetic Evidence for the Introduction of the Dingo to Australia and Implications for the Movement of People." *Journal of Archaeological Science: Reports*, http://dx.doi.org/10.1016/j.jasrep.2016.03.001.
Hamilton, A. 1972. "Aboriginal Man's Best Friend?" *Mankind* 8: 287–95.
Hamilton, A. 1982. *Nature and Nurture: Aboriginal Child-Rearing in North-Central Arnhem Land*. Canberra: AIATSIS.
Harvey, G. 2006. *Animism: Respecting the Living World*. Columbia University Press, New York.
Ingold, T. 2000. *The Perception of the Environment: Essays in Livelihood, Dwelling and Skill*. New York: Routledge.
Irvine, R. 2012. "Thinking with Animals: Horses, Dogs and Khimurr' in Eastern Mongolia." Honors thesis, University of St. Andrews.
James, B. 2009. "Time and Tide in the Crocodile Islands: Change and Continuity in Yan-Nangu Marine Identity." PhD thesis, Australian National University.
Kohn, E. 2007. "How Dogs Dream: Amazonian Natures and the Politics of Transpecies Engagement." *American Ethnologist* 34, no. 1: 3–24.
Kolig, E. 1978. "Aboriginal Dogmatics: Canines in Theory, Myth and Dogma." *Bijdragen tot de Taal-, Land-en Volkenkunde* 134: 84–115.
Lévi-Strauss, C. 1966. *The Savage Mind*. Chicago: University of Chicago Press.
Magowan, F. 2007. *Melodies of Mourning: Music and Emotion in Northern Australia*. Perth: University of Western Australia Press.
Meehan, B., R. Jones, and A. Vincent. 1999. "Gulu-kula: Dogs in Anbarra Society, Arnhem Land." *Aboriginal History* 23: 83–106.
Meggitt, M. 1965. "Australian Aborigines and Dingoes." In *Man, Culture and Animals*, ed. A. Leeds and P. Vayda, 7–26. Washington, DC: American Association for the Advancement of Science Symposium Publication.
Merlan, F. "A Mangarrayi Representational System: Environment and Cultural Symbolization in Northern Australia." *American Ethnologist* 9, no. 1: 145–66.
Nadasdy, P. 2007. "The Gift of the Animal: The Ontology of Hunting and Human-Animal Sociality." *American Ethnologist* 34, no. 1: 25–43.
Parker, M. 2006. "Bringing the Dingo Home: Discursive Representations of the Dingo by Aboriginal, Colonial and Contemporary Australians." PhD thesis, University of Tasmania, Australia.
Probyn-Rapsey, F. 2015. "Dingoes and Dog Whistling: A Cultural Politics of Race and Species in Australia." *Animal Studies Journal* 4, no. 2: 55–77.

Rose, D. B. 1992. *Dingo Makes Us Human: Life and Land in an Aboriginal Australian Culture.* Cambridge: Cambridge University Press.

Rose, D. B. 2011. *Wild Dog Dreaming: Love and Extinction.* Charlottesville: University of Virginia Press.

Ryan, J. S. 1964. "Plotting an Isogloss: The location and Types of Aboriginal Names for Native Dog in New South Wales." *Oceania* 35: 111–23.

Smith, B. P., and C. A. Litchfield. 2009. "A Review of the Relationship between Indigenous Australians, Dingoes (*Canis dingo*) and Domestic Dogs (*Canis familiaris*)." *Anthrozoös* 22, no. 2: 111–28.

Smith, B. P., and C. A. Litchfield. 2010. "Dingoes (*Canis dingo*) Can Use Human Social Cues to Locate Hidden Food." *Animal Cognition* 13: 367–76.

Thomson, D. F. 1949. "Arnhem Land: Explorations among an Unknown People." *Geographical Journal* 114: 29–43.

Thomson, D. F. 1957. "Yellow Dog Dingo." *Walkabout.* May 1, 16–18.

Thomson, D. F. 1972. *Kinship and Behaviour in North Queensland: A Preliminary Account of Kinship and Social Organisation on Cape York Peninsula.* Australian Aboriginal Studies No. 51. Canberra, AIATSIS.

Trut, L. N. 1999. "Early Canid Domestication: The Farm Fox Experiment." *American Scientist* 87, no. 2: 11.

Warner, W. L. 1964. *A Black Civilization: A Social Study of an Australian Tribe.* New York: Harper and Row.

White, I. M. 1972. "Hunting Dogs at Yalata." *Mankind* 8: 201–5.

Wilkins, A. S., R. W. Wrangham, and W. T. Fitch. 2014. "The 'Domestication Syndrome' in Mammals: A Unified Explanation Based on Neural Crest Cell Behavior and Genetics." *Genetics* 197: 795–808.

4

FARM ANIMALS IN A WELFARE STATE

Commercial Pigs in Denmark · *Inger Anneberg and Mette Vaarst*

Can the concept of "animal welfare" be helpful as a lens through which "domestication" meaningfully can be viewed and alternatives to classical views of domestication can emerge and be discussed? Can this open the way for seeing the pigs differently, or seeing them at all? Or does this way of envisioning the pigs create new moral and ethical dilemmas in the name of the "animal welfare state"? These are the main issues to be discussed in this chapter about pigs on commercial large-scale pig farms in Denmark. Before we look into this, we need to describe the industry as it operates today.

"Danish bacon" and Danish pig exports are well-known far outside Denmark's borders (Hamann 2006), and guests from abroad sometimes express surprise that pigs are not more visible in a country with 12.8 million living pigs and only 5 million citizens.[1]

However, today, pigs in Denmark are kept mainly indoors, just as nonorganic dairy cows and chickens are.[2] Still, it is possible to find in the landscape evidence that Denmark is a country with a huge pig production. The first sign might be the intensive monocultural fields of barley and wheat, colza and corn: 80 percent of Danish farmland is used for growing fodder for farm animals.[3] Characteristics specific to pig farms might be two parallel, long and low concrete farm buildings with small windows and high fodder silos next to them, and one or two low, round slurry tanks covered with a gray, pointed rubber shield. In spring and autumn, the wind from the countryside periodically brings the harsh smell of ammonia, both day and night, when machines spread slurry on the fields as fertilizer.

Getting access to one of the 3,800 pig farms in Denmark is not a simple task.[4] There are signs at some barn doors restricting access, and there are other sorts of boundaries—like a twenty-four-to-forty-eight-hour quarantine between visits to farms. This is due to food safety regulations and fear of contamination, mainly between the pig farms. Pig farms are categorized according to disease status, namely, which infectious diseases have been eradicated from the farm. In the past few years, a dramatic rise in antibiotic-resistant bacteria (CC398-MRSA) in pig herds, infecting pigs and the humans who work with them, has exacerbated the fear of contact between pig farms and their surroundings.[5] As a consequence, school classes for a period were not allowed to visit Danish pig farms (Politiko.dk 2014). All this contributes to the continued invisibility of pigs as animals in the Danish landscape.

Behind Closed Doors

If you *do* get access to a farm with animals, what is it like inside? Today almost all industrial animal farms in Denmark are specialized, having either only dairy cows or only pigs.[6] The pig farms are also specialized. Some have merely one part of the pig production, for instance sows and piglets up to an age of between three and four weeks (when they are weaned) or only slaughter pigs.[7] Some have the "full line," meaning the whole pig production cycle, from farrowing sows to slaughter pigs.

Inside the barns, you will find interconnected sections, with, for example, one section for service (insemination of sows), one for pregnant sows, and one for farrowing sows with piglets.[8] After weaning, the piglets will be taken to a separate building with strict regulation of temperature and feed. Different facilities give very different impressions and experiences, depending on whether the building is new or old and has more or less ventilation. The air and smell can also depend on, for instance, whether straw is used and on how clean the place is kept.[9] Sometimes old, dirty windows draped with spiderwebs give a dull, unpleasant impression; clean windows with more daylight coming in make the place seem brighter. Even the sound from the pigs can differ widely from place to place and between sections, depending mainly on the time of day and what is going on around the pigs. Just before feeding the noise can be incredible. When castration and docking is done in the farrowing section, piglets scream in panic and pain, and when new groups of sows are establishing their hierarchy, the fights among them can be noisy. Otherwise, after being fed, when the pigs are resting and no mutilation is going on, a pig barn can be very quiet, though regular grunts of communication will be heard, for instance, in the farrowing unit when the sow grunts to signal to the piglets that milk is being laid down and the piglets squeal while finding their favorite teats for sucking.

Is This Classic Domestication?

In this chapter we focus on the Danish pig industry, and as a starting point we argue that the industrial setting of Danish pigs illustrates in the most obvious ways Juliet Clutton-Brock's concept of domestication in anthropology, wherein domesticated animals are defined as "bred in captivity for purposes of economic profit to a human community that maintains complete mastery over its breeding organizations of territory and food supply" (Clutton-Brock 1989, 7). As stated by Rebecca Cassidy (Cassidy 2007, 5) in *Where the Wild Things Are Now,* this definition was primarily economic and biological and based on the incorporation of animals into exploitative relations with humans. This description is consistent with the fact that Danish pig producers historically have had a high degree of control over the breeding of pigs (Fredeen and Jonsson 1957), which is part of the narrative of success often told about the industry. Thanks to the industry's constant focus on genetic development, sows now give birth to a lot more piglets than a sow would be able to give birth to or bring up under natural conditions.[10] A narrow focus on genetics has produced pigs that grow fast, have a very high feed efficiency, and become lean slaughter pigs with little fat.

Nevertheless, this very industrial setting, which has become the platform of Danish pig domestication, with its focus on creating fast-growing, uniform "animal machines," has faced new challenges in recent years, in particular those related to the notion of "animal welfare." This has become an everyday phrase in Denmark, even though we do not have a common definition of what animal welfare means.

We will argue that although the economic and biological definition in one way fits Danish pigs, formed and altered as they are by genetic selection, fodder, and effective indoor systems, the term "animal welfare" has opened new possibilities for seeing the pigs differently—not just in practices formed by domestication but also in practices connected to "animal welfare," which here means not only legal guidelines but also the day-to-day interactions between pigs and farmers. Using "animal welfare" as a lens gives us the opportunity to discuss whether this term offers humans and nonhumans a different relational agency than the classic theory of domestication, with its focus on human mastery. This also paves the way for a discussion of the birth of "the animal welfare state" and the moral and political dilemmas in the present animal welfare legislation. To discuss this, we turn toward ethnography and move into the barns.

CASE 1: A CONSTANT STREAM OF INTERVENTIONS

Anton, at the time of the study, works as a manager on a pig farm in Mid-Jutland, Denmark, and finds it complicated to work with pregnant sows in groups—instead of isolating the sows in crates—even though he has done this since 2009.[11] *However, he also says that he would not do without it. He says, "After being loose, the sows, they seem to get stronger legs and also manage to farrow a lot easier than the confined sows."*

Anton expresses both satisfaction and frustration with this system for the sows, when Inger (first author) follows him in his daily work on the farm in connection with her ethnographic fieldwork on Danish farms.[12] *In this pig production, Anton is in charge of eight hundred sows, and these sows are loose in groups of between thirty and thirty-five during pregnancy until one week before farrowing, when they will again be confined to their farrowing pens. Anton, while looking at the sows, explains various methods that he finds important but also poses a challenge related to the group-housed sows in daily practice:*

> *You have to establish the groups in the right way, because the sows will always try to form a hierarchy, and if the groups are unstable or mixed in a way that is new to the sows, they will fight more and injure each other. Furthermore, the sows are fed on the floor in this system, and some of the thin sows cannot get sufficient access to fodder. If such a fat lady becomes completely dominant in a pen and eats too much, I sometimes take the thinner sows out and confine them again in a crate, although the legislation does not allow that, to give them more fodder.*

If the sows do not have enough space, they tend to fight, especially if they lack straw. In Anton's pigpens, straw was not used, because the slurry system could not cope with it.

He says that the sows get frustrated if they cannot find a place to rest, or if they have to rest in a dirty area, which they do not like. But they also react to the people who look after them: "You can feel immediately if they are being kicked by a worker. They become aggressive—they become like the aggressive human who harmed them."

Anton does not mind legislation that liberates sows from their crates, but he complains about the constant stream of new interventions from the authorities: "It makes me angry that they for instance keep changing what size the pens should be. It's expensive for the farmer with all these changes, and if the farmer gets financial problems, there is no job for me."

This excerpt suggests that engaging in daily life with animals in an intensive industrial setting constitutes something different than a classic case of domestication. Anton is not an almighty sovereign master over his domesticated pigs. His experiences of domestication are as much about a lack of control as about the acquisition of it.

In other words, Anton's experience dismantles the narrative we normally live by: the strong narrative of domestication that tends to define animals through idioms of purposeful human mastery and to emphasize control as a characteristic feature of the human-animal relationship (Law and Lien 2014). State legislation is rarely part of such stories. Anton's story also demonstrates that he sees and perceives his sows in a different way when they are loose than when they are tethered or confined. He recognizes their need, for instance, to establish a hierarchy and to use different zones in this common pen for different purposes: one for defecating, one for eating, and one for resting. In this way, the story shows how the legislation and control that introduced elements into Anton's farm aimed at better meeting the animals' needs also made Anton see them differently. He became "domesticated" by legislation, which changed his perception of and relations with his pigs.

We propose thinking about animal welfare legislation not only as a domestication of farmer-pig relations but also as a domestication of ways of doing domestication well. As Lien (2015) writes of welfare legislation concerning fish, "such legal guidelines appear both too idealistic and at the same time insufficient, as no checklist can ever completely remove the moral practical dilemmas involved in raising farm animals" (Lien 2015, 142). Following this, Anton's case also shows that animal welfare legislation can silence the farm owner or the workers and make them feel unmotivated, discouraging them from taking initiatives and at the same time restricting their daily practice with new expenses and legal requirements.

Furthermore, the legislation always excludes or expels the most appropriate animal welfare for the time being, responding instead to requirements from the market and from the different political systems. From this angle, legal animal welfare guidelines be understood not as something that provides better conditions or even agency for farm animals but rather as another variation of an asymmetric relationship that is practiced, promoted, and enforced by the country's legal authorities. In "Good farming—Control or Care?" Singleton (2010) discusses whether quality can be ensured by introducing systems of control in farming, in this case ear-tagging. A farmer describes how she feels about having to follow the requirement from the authorities to ear-tag the cows: "Keeping the livestock alive is one thing. Keeping on the right side of DEFRA, the environment agency and the taxman, that is another" (Singleton 2010, 236).

This situation described by Singleton shows how "care" and "control" are perceived as two distinct aspects of having animals. However, farming practice cannot be seen as simply control, because it involves humans and animals living together adaptively (2010). We will argue that the legislation can also be seen as a cross-cutting between care and control; in this case, it could be argued that the ear-tagging of cows helps the farmer to identify and therefore follow each individual animal.

In the next part, we look more deeply into how animal welfare legislation has developed historically. Then we will show through ethnography how animal welfare legislation is encountered by inspectors, farm owners, and farm workers in daily practice, and we will discuss whether domestication by the welfare state raises new possibilities of agency for farm animals and human animals.

Welfare State Domestication

To get closer to the birth of the animal welfare state, one first has to look at what it is like to be an industrial animal in, for instance, Denmark. According to Harfeld (2010), there is no academic consensus as to the exact characteristics of *industrial agriculture*, but several supplemental and to a certain extent overlapping characteristics are usually employed (2010, 104). Harfeld points with Fraser (2008a, 167–89) to *intensification*, for instance, of systems that house more animals in fewer square meters, of increased indoor housing systems, and of feed efficiency.

Industrial intensification also relates to the amount of output—for example, more muscle mass and quicker growth in broilers and pigs, more milk from cows, more eggs from laying hens, more piglets from sows—per input unit, typically a person hour. In addition, it applies to monoculture farming, where only one animal species is kept, in many cases animals of only one age group or during a particular phase of the animal's life, and it applies to farm ownership, with increasingly larger farms owned by a few farmers. Finally, Harfeld (2010) also notes that industrial intensification relies on several scientific innovations, for instance, such medical contributions as the prolific use of antibiotics, vaccines, and hormones, and also the technological advances in machinery that replace human labor and reduce the role of human knowledge as well as dramatically reducing the number of minutes the caretaker spends on each animal.

In the periods after the First and Second World Wars, Europeans seemed occupied with human survival and meeting their basic needs. As Fraser states (2008b, 61), animal issues "were largely consigned to a back burner." But in the 1950s and 1960s, when security and affluence returned in the industrialized nations, society began to pay increasing attention to issues that went beyond survival and personal prosperity. In 1964, *Animal Machines: The New Factory Farming Industry* (Harrison) was published in Great Britain. The author, Ruth Harrison, a layperson, had set out to study what had happened to farm animals in industrial agriculture in her country. A vegetarian, Quaker, and conscientious objector, Harrison set out to tell the public how veal calves were raised and how hens were being kept in tiers of cages. In the process, she revealed how farming had changed from an agrarian to an industrial activity. Fighting for these mistreated farm animals became her lifelong calling. She wrote in the foreword to her book: "I am going to discuss a new type of farming, of production line methods applied to the rearing of animals, of animals living out their lives in darkness and immobility without a sight of the sun, of a generation of men who see in the animal they rear only its conversion factor into human food."

The concept "animal welfare" had not yet been invented when Ruth Harrison wrote her book. She does not use the word "welfare" or "domestication" either but instead employs such words and phrases as "factory farming," "cruelty," "exploitation," "our domination of the animal world," "to rob them of all pleasure in life," and "to treat living creatures solely as food converting machines" (Harrison 1964, 3).

Legislation from this period and earlier was aimed at preventing overt cruelty toward animals, but it did not focus on animals as sentient beings who responded to their surroundings or had emotions. Ruth Harrison writes in general about the legislation at that time that it was "loose and ill-defined" and, moreover, "hopelessly out of date." In his *Understanding Animal Welfare: The Science in Its Cultural Context*, David Fraser (2008a, 66) writes that when Ruth Harrison's critique appeared "there was no established meaning of the term 'welfare' when applied to animals. Hence, the furor surrounding *Animal Machines* also stimulated a remarkable debate on an issue that had never really been settled: What actually is animal welfare? What constitutes a good life for animals?"

Harrison introduced several themes, which are still part of the present debate, including "animal suffering" and "the unnaturalness of the various production systems," or, in Harrison's words, "animals being deprived of 'all pleasures of life.'" In Britain, the book led to the appointment of a committee to investigate the welfare of animals kept under intensive livestock husbandry systems. One of the important recommendations in the resulting Brambell Report was the call for research in veterinary medicine, stress physiology, animal science, and animal behavior.[13] Along with encouraging the development of a whole new "animal welfare science" for farm animals, the Brambell Report also led to the development of legislation and the setting of minimum standards for the treatment of farm animals, first nationally and eventually by the European Union. The first European regulations to protect animals were adopted ten years after the report was released and were consistent with the conclusions of the report, which was influential not only in the United Kingdom but also in the rest of Europe (Veissier et al. 2008).

Animal Welfare in a Welfare State

In Denmark, there is a convergence between the influence from the Brambell Report and what is sometimes called "the golden age for the European and Danish welfare state," the years between 1950 and 1970. A universal approach was taken to human welfare; for instance, state pensions for everybody were implemented in 1956 (Ploug 2004). Anneberg and Bubandt (2015) point out not only that the Brambell Report coincides historically with the consolidation of the welfare state but also that its theoretical focus argues for a concept of state-governed welfare for animals that is closely connected to a central idea in the political philosophy of human welfare—namely, the understanding of freedom. The Brambell Report talks about "the five freedoms" for animals, and the infusion of freedom for production animals can be seen as consistent with Michel Foucault's (1982) argument that freedom in modern politics is not a threat to power but a tool that can be used to govern people.[14] Anneberg and Bubandt (2015) argue that with the Brambell Report, governmentality has been introduced into the barns. Animals should be "given freedom"—however, this freedom is framed by lots of juridical restrictions and is mostly expressed as "freedom from" rather than "freedom to," which can make a big difference, as we soon will see. Farmers lost freedom when they were restricted by legislation (as expressed by Anton in the case above), but they were at the same time offered the possibility of seeing the domesticated animal in a new light.

Anton represents a duality in the farmers' view when it comes to acceptance of the constant influence of the state in their daily routines: he accepts it, but he also protests what he perceives as the totalitarian state interfering with his daily work with his animals.

As shown by Anneberg et al. (2012), Danish farmers, on the one hand, agree that state enforcement of animal welfare legislation is necessary, because they fear that otherwise their colleagues will cheat. But, at the same time, they oppose being constantly told "from above" how their animals should be kept, and they feel overwhelmed by the amount of legislation on animal welfare. Farmers ask for fairness and consistency from the inspectors judging the animal welfare on their farms, but, at the same time, they want their farm to be judged on a case-specific basis, considering its unique context (Anneberg et al. 2012).

Domestication via Welfare

The duality given in the example above, accepting and protesting at the same time, is consistent with how domestication historically is described as a project of the state. James Scott wonders at this when he reflects about why we "as species, having spent virtually all our span on planet earth as hunters and gatherers, ended up assembled in great clumps, growing grain, tending livestock and governed by the political units we call states and empires?" (2011, 186). But neither states nor the forms of domestication with which they are intertwined are generic. Both can be related to how the state has intervened in agriculture in Danish history and how "the good life" is understood in a Danish state/human context. In Denmark today, domestication can certainly be recognized as part of a project through which the state has worked with the farmers' organizations for a long time, for instance when promoting bacon and butter (Higgins and Mordhorst 2008). Thus, Anton and his predecessors have been accustomed to state interference, but in relation more to export and marketing than to the rights of the animals.

However, trust in the Danish welfare state—that the state generally guarantees the good life—has been shown to be a fundamental part of how people perceive themselves as citizens in Denmark. Jenkins (2012) argues that

Danes in general accept a high degree of detailed legislation and control, because the state also more or less consistently delivers the goods of a modern welfare system, which Danish citizens are brought up to see themselves as part of: "Furthermore, the central state is not only generally unobtrusive, much like a neighbour—the state-next-door, if you like—but it is also generally a good neighbour" (2012, 169). Inspection of animal welfare is consistent with a daily stream of other state interference that comes along with the welfare state, which farmers also are aware of, because they know it from friends and family. As stated by a farmer in Anneberg et al. (2012), "In a home for old people, we know that they [the staff] use so much time filling out documents instead of looking after the old people. . . . This is not good care . . . We are farmers because we care for animals, not because we are interested in control or filling out papers."

The threat of sanctions from the authorities has become a part of the Danish pig farmer's life, and in the context of the Danish welfare state, sanctions are often combined with offers of guidance, on how to become better humans. Contemporary welfare policies embrace liberal ideas of freedom by emphasizing the need to reduce state control and expenditure and for institutions and citizens to take more responsibility for themselves (Mik-Meyer & Villadsen 2013). There is a constant negotiation between citizens and authorities in handling the paradox of "too much or too little," or as Mik-Meyer and Villadsen put it, noting that it is an ambition of the welfare state to treat each citizen as unique: "However, these institutions are characterized by distinct professional logics or rationalities, which guides how the staff observe and encounter the citizen. . . . The ideal citizen is one who acts responsibly, is strong-willed and acknowledges that he or she plays the essential role in solving his or her own problems" (2013, 4).

Negotiations about animal welfare have in this way been incorporated into the welfare state, because here it is obvious that a discursive and practical framework shapes the encounter between the citizen, the professional, and the animal. Farmers know, for instance, that starting an argument with the inspector might make things worse. Some even strategize about how they want to be seen when they meet the inspector: "Spending time grumbling about negative things is a special farmer attitude that I don't like. . . . You should always be open to a dialogue with the inspectors, invite them in for coffee and talk to them. This is much more effective" (Anneberg et al. 2012).

In the following case, we illustrate such negotiations between authorities and farmworkers and what these negotiations actually achieve on behalf of the animals.

CASE 2: INTERPRETING FARM ANIMALS: GIVING THEM DIFFERENT AGENCY

Inger (author) follows a group of inspectors on a visit to four different Danish pig farms, where they look for compliance and noncompliance with the EU legislation on pig welfare. The inspectors must discuss and agree on whether the farms are in compliance with the legislation. On some farms, problems are found: The loose-housed pregnant sows seem to prefer to stay inside their crates (where they also are fed) instead of going out into the common area, where they can be loose, root, and be social. Only one or two dominant big sows lie in the common area.

At one farm, an inspector speaks with a farmworker, asking: "Do the sows always, all day long, stay in their crates or is it just now, when we are here? When do they come out?"

The farmworker answers: "They never come out. It's OK. They prefer it like this, to be in their crates. It's like before when they were confined. They do not go out there even if they can. And we cannot offer more straw as this is what our slurry system can manage."

The inspector does not comment further during this visit, but when the inspectors come together to discuss compliance and noncompliance, they note that even though the common room for the sows did fulfill the demands of the legislation, being of the required physical size, the space outside the crates was not at all attractive for loose sows. There was very little straw out there, nothing that would meet the sows' motivation for rooting—and no walls to hide behind if a dominant sow came out. The area where they could lie down was wet; the sows could not make different zones for defecating and resting, which is what they prefer. Also, the inspectors found that the behavior of these sows could be interpreted as their being constantly hungry, so this could have been another reason for them to stay inside the crate: they hoped to get food.

Some of the inspectors were frustrated during the discussion. Having learned that sows have a basic need to root and socialize, they had just found out that this need could not be fulfilled in the system legalized by the EU. On some farms, farmers and farmworkers agreed to interpret the behavior of the sows as "sows prefer to stay in their crates," while some of the inspectors interpreted it as the need to keep the sows under control.

A paradox related to the needs of the loose-housed sows becomes clear. Farmworkers in this case found that the sows do have agency—and they interpret it like this: The sows can choose to stay in their crates. They are given a choice to be social, but they prefer not to be.

The inspectors also found that the sows have agency but concluded that it was *impossible for the sows to realize their potential*, although the requirements of the legislation had been satisfied. There is no alignment between the subjects that the state inspectors think pigs are and the subjects that farmers think pigs are. We see this as an example of an attempt to reshape farmer subjectivities through the legislation and thereby encourage them to see pig subjectivities differently. In this way legislation related to animal "welfare" always balances between trying to improve the animals' lives and trying only to "mitigate suffering." In the next section, we give examples of this compromise.

Animal Welfare Legislation and Animal Welfare Science: A Compromise

In anthropology, legislation has been described as a very unsteady instrument, and animal welfare legislation fits this description well: sometimes it is described as a *compromise between many different interests*, with welfare being just one of them (Forkman 2011). Lien (2015) also stresses this aspect when she shows how animal welfare legislation is no longer restricted to four- and two-legged, furry and feathered companions. Farmed fish have emerged as sentient beings as well and thus are subject to welfare regulation on the assumption that fish such as salmon *might* have the ability to feel pain (2015, 126). But Lien stresses that while both philosophical and biologically grounded arguments in support of salmon sentience undoubtedly have made a difference in specific legal situations, they often fail to consider the heterogeneous relations through which farmed salmon come into being—the lively practices of growing salmon in tanks and pens (2015, 127).

The legislation related to pig welfare in Denmark also can be described as failing to consider the practices in the pens. An example is the castration of pigs, which is a painful surgical procedure performed without anesthesia in most EU countries, reducing the castrated pig's welfare significantly compared to that of the uncastrated pig. According to the Danish animal welfare act, paragraphs one and two, *animals should be treated carefully and properly and protected as well as possible from pain, suffering, fear, disablement and major inconvenience.* However, castration without anesthesia is still permitted according to Danish legislation, though now it is permitted only if the procedure takes place as early as possible, and within the first two to seven days of the piglet's life, and if pain relief is administered. Many farmers defend this method, even though they say at the same time that they would prefer not to do it at all. In their daily practices in the barns it's about effectiveness, and farmers find giving anesthesia or even painkillers to be time-consuming (Tuyttens et al. 2012). Another example of a compromise in legislation is the mandate that sows must have access to sufficient nesting material in the last week before they are expected to farrow *unless* this is "technically impossible to handle"—a reference to a farmer's slurry system. In practice, it seems to be "technically impossible to handle" in the majority of cases. A third example is the tail docking of pigs. Routine tail-docking of pigs is illegal in the EU (Statuary Order 324, paragraph 4). However, tail-docking of piglets is allowed within the first two to three days of the piglets' lives, if a farmer can document that tail injuries in the herd are attributable to the omission of tail-docking. More than 95 percent of pigs are still having their tails docked in the EU, and thus the legislation is ambiguous, on the one hand, forbidding tail-docking and, on the other hand, creating a possibility that the exception becomes the reality.[15] One paradox is that tail-docking is accepted even when the piglets are sold to an "unknown receiver," where of course no documentation of potential causes of tail injuries can be presented.

Thus, animal welfare legislation can be understood as, in one way, giving agency to animals and also as silencing both animals and farmers by functioning, as Michel Callon puts it, "to silence those in whose name one speaks" (Callon 1986, 14). In this case, the legislation explicitly gives "easy escapes" by offering reasons to avoid following the legislation, such as "tail should not be cut unless it is technically impossible to handle pigs with tails." One could ask whether and how the legislation is being "domesticated" by the political power of farmer organizations and the agriculture industry, and whether the relationship between the farmer and the state/legislation is one of mutual shaping—almost like codomestication? The legislation allows practices that restrict the animals while ostensibly protecting the animals from suffering, hence convincing the surrounding society that "things are taken care of."

The animal welfare science that followed the Brambell Report goes hand in hand with the development of the law when it is understood as an unsteady instrument. According to Fraser (2008a:6), animal welfare science came as a response to the ethical concerns about the treatment of animals and the debate about the kind of life they should be allowed to live. Animal welfare science rose as a "mandated field," which refers to science that is done for a social purpose, for example, to guide policy or legislation—a science that has emerged as a means to provide answers. The rise of animal welfare science seems linked in part to the classical understanding of domestication as total human control over animals. See, for instance, the American scientist McGlone (1993), who developed a theoretical framework to distinguish between acceptable and unacceptable welfare. In his paper "What Is Animal Welfare?" he wrote that because "feeling a little poorly is much like feeling hungry (something we all normally experience from time to time), this cannot be the critical measures of well-being. Rather, only when animals reach a pre-pathological state can we say welfare is poor" (McGlone 1993, 26). In this understanding, an animal must be nearly dead to be seen as "in a poor state," a definition that could hardly be more asymmetrical: human animals defining when other animals feel poorly.

Today there seems to be agreement within animal science that there is not one shared definition of animal welfare. Three overlapping approaches are often described: the natural, the emotional, and the physical. Often, scientific research finds that consumers prefer naturalness, for example, that cows have access to pasture or sows opportunity to root outdoors, while farmers prefer the physical approach, linking welfare to, for instance, health. The third approach, the emotional (sometimes also called the affective), has been found to be preferred by animal scientists and by the authorities (Sorensen & Fraser 2010).

The balance between the different definitions can be seen as a political issue, but it is also a constant part of an animal scientist's work. Some scientists might show that industrial pigs at a minimum ought to have access to straw to fulfill their need for rooting or sows' need to build a nest. Other scientists might reach different conclusions about how much or how little straw is needed. Animal science, therefore, involves heterogeneous relational practices. Politicians, actors from different organizations, and the farmers with their day-to-day practices all give different arguments for why they cannot or will not implement the legislation or agree on the results found by science.

However, as part of the growing understanding of animal welfare as a complex field wherein compromises in legislation restrict animals, some scientists have taken a different approach. They emphasize that when defining animal welfare, we must assess "the whole animal" and describe the expressive qualities of its demeanor (e.g., anxiety, contentment). Such indicators must be validated in the same way as other indicators of health and behavior (Wemelsfelder and Mullan 2014). Wemelsfelder and Mullan (2014) stress that developing animal welfare indicators is a complex task and also that the use of the indicators ultimately depends on how humans engage with animals in daily life, depending on their skills, attitudes, and efforts. They conclude: "We are used to framing that engagement in terms of control, but increasingly academics across disciplines are encouraging us to think more inclusively of animals as members of "more-than-human communities." Thus, both new knowledge of animals and new ways of relating to them are needed to make practical welfare assessment tools a success. Monitoring welfare standards is an important first step, but only a greater sensitivity and responsibility towards animals will make these standards work" (Wemelsfelder & Mullan, 2014, 116). The animal welfare science in the above example—wanting to think of animals as members of "more-than-human communities"—emphasizes management as knowledge, skills, and relationship rather than as control. This is in line with developments in anthropology, where domestication can be viewed as the outcome of a series of complex relationships between humans and other animals in different places and at different times (Hurn 2012, 64). However, freeing themselves and their animals from legislation and the control of animal welfare is not a possibility for Danish farmers. Of course, they can choose not to follow the law strictly in daily life, but today they are watched closely, not only by the authorities but also by the surrounding society, and they are always at risk of being inspected by the Danish Veterinary Council. Nevertheless, it *is* possible to take one's farm system and one's animals further than the legislation demands, as our last case will show.

CASE 3: BEYOND LEGISLATION: BEING FACED BY THE SOW

It's August 2014, the harvest is over, and the farmer has time for visitors. We are a group of scientists from Animal Science, Aarhus University, Denmark, who have gone to visit a conventional pig farm in eastern Denmark.[16] *The purpose is to see a brand-new pig facility, opened in spring 2014 and built to house the farmers' 1,000 farrowing sows. It is constructed following a new design in which farrowing sows are not confined but loose, each in their individual indoor farrowing pen together with her piglets. Only five farms in Denmark (out of 3,600) have this system; the rest keep their sows in crates during farrowing and only have pregnant sows loose in groups until they farrow, because this is what the EU animal welfare legislation has demanded since 2013.*[17]

While planning and building this new farm, the farm owners (father and son) decided to anticipate what they expected to be future legal requirements: that the farrowing sows must be loose while having their piglets and during the nursing period.[18]

When showing us the new section, the son enthusiastically tells us about this system, which is very different from the systems of his earlier experience. He stressed that it has changed his view of these animals:

> *They are different animals when set free. They behave differently and we have to learn to look at them in a new way, have to treat them as individuals. We cannot just do the automatic treatment with castration and tail-cutting of the piglets as we did before, when the sow was in the crate and she could not move around, when we came into the pen to handle the piglets. Now we have to look at the sow and let the entrance to the pen depend on*

her mood and attitude, we can't just push on. You can say that the animal has become more visible, and we know she is in charge in the pen. Before, when confined in a crate, we could also feel she was frustrated, but she could not act. But now she can face us, look at us.

The biggest skepticism and resistance to this system we have seen from our farmer colleagues. Some fear the loose farrowing sows, expecting them to be violent and much more difficult to look after. Stories tell about these sows attacking the farmworkers. But we have not seen anything of this at all. Also mortality rate is constantly put forward by other farmers—the sow lying down on the piglets and killing them if she is free, but this hasn't been a problem either. Not yet . . .

One of the scientists asks the farmers whether they can get extra money for their pigs now that the farrowing sows are loose. The farmer answers: "We have had interest from one of the big consumer chains for our meat, but they raised questions because we don't have "the right animal welfare" in all our sections—for instance, we still cut the tails and we castrate. As long as we only have part of the welfare in the way they demand, they will not accept our meat as "welfare pigs" and not pay us any extra." Using his experience is central for this farm owner, and it shows how sentience is specific and relational. He does not see himself as in opposition to legislation, but in a follow-up interview with the first author of this chapter, he says that he is motivated because, through his management skills, he is generating changes for his sows. He uses expressions like "train the animals to understand that it is OK when we enter the pen," and "teach the farmworkers how to behave calmly," and "be much more aware of the sow than we were before." Nevertheless, he also expresses a desire to change the sows genetically, to breed sows that are less aggressive will be a sales parameter in the future, as he sees it.

He emphasizes that the younger generation of farmers will be more focused on creating changes for the animals: "Not that we are passionate animal fanatics. We have to get used to these new thoughts ourselves. It's a long process, but today the pleasure of entering the farrowing barns is important for me. It is a great part of it all. Anyway, I still think that these changes will occur by themselves, because we are a new generation."

This shows the complex motivations both of wanting to change things to create something better and more future-directed and, at the same time, of being aware that these changes may also be required by future legislation. These farmers, father and son, went beyond the minimum requirements of the legislation in planning their new farm, and the changes they implemented made the animals' own agency more visible. They explain this in a language that does not connect to numbers, although turning pigs into numbers proved a crucial feature of today's pig industry, in order to track production efficiency, among other things (Anneberg, Vaarst & Bubandt 2013). The sows will still be counted, changed into numbers—such as the number of piglets born to a sow per year, the number of piglets that lived, and so on—because electronic control ("e-control," which is standard in the Danish pig industry) exists at this farm also. But in addition, the sows have got a "face turned towards the farmer," and the farmer "feels pleasure" entering the pens, acknowledging that the sows behave differently than before.

Conclusions and Perspectives: Hope or Willed Blindness?

As mentioned earlier, the Danish pig industry, with its constant focus on genetically shaped pigs, formed and altered inside the barns, in one way clearly embodies Clutton-Brock's definition of human mastery over nonhumans. At the same time, these massive lenses of domestication as "pure control," with their emphasis on profitability, have clear limitations for giving agency to pigs. The "successful narrative" of human mastery has partly made it possible to ignore the fact that farm animals have for years responded fiercely to the circumstances under which they are kept: the pigs in intensive production systems get sick and are massively treated with medicine; they get harmed, systematically mutilated; and they are at risk of dying because of this form of domestication. The animals are now shaping legislation and the state; in other words, they shape the conditions of their own domestication. Being confined as farrowing sows without the ability to turn around and then being brought back into heat as fast as possible to produce more piglets carries the price of a very short, often painful life, a life full of frustrations.

This is why we have brought the concept of animal welfare into our story. Although massive state control is influencing this concept, and although we have illustrated the many compromises "animal welfare legislation" is built on, we find that it does offer us a "different lens," which includes the potential to give animals agency and provide an alternative view of animal farming not only as an industry aimed at making a profit but also as a way of interacting with living sentient beings. Thus, legislation about animal welfare is not to be positioned merely as another aspect of domestication via human mastery. Legislation as part of domestication has both limitations and possibilities for farmers and pigs and their relations with each other and the state. Both farmers and pigs are shaped by and contribute to shaping animal welfare regulation. Lien (2015) approaches the "newcomer to the farm," the farmed salmon, as a relational quality, a potential aspect of heterogeneous relational practices, and she suggests that it is in the space of uncertainty that the potential for improvement lies.

Sometimes knowledge about animal welfare is introduced to farmers by the inspectors who monitor the enforcement of animal welfare legislation. This can be a negative aspect, as discussed by Singleton and also illustrated in our own case with Anton. There is a risk that governmental control can silence the farmers, by lifting their feeling of responsibility, and that instead of talking about the relational aspects between human and nonhumans we talk about the relation between the state and the individual human. When we focus on the human mastery aspect of domestication, we risk overlooking the fact that farm animals actually do "talk back," even though humans tend to read their language through the lens of our own agendas. But in the third case, with the move toward loose farrowing sows in anticipation of legislation to come, we have shown how legislation can also encourage new ways for pigs and people to be together.

Here it is relevant to ask: What if it were simply left to consumers to buy products that meet their expectations? Their choices would depend on what they knew and saw regarding farming. As we illustrated above, the landscape of a confined animal industry does not allow Danish consumers to see how animals live. The state has the potential to speak on behalf of not only "consumers" but also citizens, both animal and human citizens, and, in this way, to give agency to the actors without a voice—the animals—and to inspect what is happening inside the places where they are confined. Still, states do produce and legalize a form of domestication, as described by James Scott and others. State domestication can provide reasons for resistance to farmers who are in a price war with their competitors and see legislation and inspection as restrictive. It is a challenge for scientists, farmers, opinion makers, politicians, and NGOs alike *not* to become caught up either in explaining "animal welfare" only as "meeting the nature of animals" or in using it as a weapon in an ongoing war and in whatever direction the actor wants to influence the debate, the legislation, or the development of animal farming.

Later in this volume, Anna Tsing suggests that what she names "unintentional cultivation and domestication as re-wilding" can offer a hopeful alternative to imagining multispecies life with humans as components. She asks: "What if we redefine the 'homes' we wanted to make with other creatures to include the worlds they make as well? Might we come up with better ideas than industrial ruins?" In this chapter, we have discussed and offered a very small and uncertain alternative to the industrial ruins left by the pig industry: the possibilities that lie within animal welfare legislation. Perhaps we are too blind? Perhaps we, as inhabitants in a massive industrial farm setting, have become victims of what Gjerris (2012 and 2015) calls "the willed blindness of humans"—wherein we seem to live in a "willed blindness" to the effects that our meat production and consumption have on animals, the environment, and the climate (Gjerris 2012, 35).

Gjerris argues that this willed blindness cannot be explained either by lack of knowledge or by scientific uncertainty; he suggests that the solution lies in a new moral vision of our obligations, a *new vision of what a good life is.* As we have seen, some Danish farmers argue that seeing pigs differently, as living sentient beings, is possible, but it takes time for this point to be reached. New "homes" must be built for them, and new relations between farmer and pig created, before we can respect what it is like to be a pig.

NOTES

1. Statistic Denmark, accessed March 24, 2015, http://www.dst.dk/da/Statistik/NytHtml.aspx?cid=18830. Of the 12.8 million pigs, one million sows each year produce 29 million pigs. 90 percent of the 29 million pigs are exported. The agricultural organizations estimate that exports of pig meat account for almost half of all agricultural exports and for more than 5 percent of Denmark's total exports. Accessed May 5, 2015, http://www.agricultureandfood.dk/Danish_Agriculture_and_Food/Danish_pig_meat_industry.aspx.

2. In Denmark, 12.3 percent of the pig farms and 10.6 percent of the farms with dairy cows are organic, meaning that among other requirements these animals must have access to outdoor areas (https://www.landbrugsinfo.dk/Oekologi/Sider/danskoeko.aspx#Økologisk landbrug i dk). The nonorganic animals do not by law have the same rights or possibilities.

3. Denmark comprises 4,310,000 hectares of land, of which 62 percent is cultivated as farmland (Johannsen et al. 2013; Natur- og Landsbrugskommissionen 2012), making Denmark one of the most intensively cultivated countries in the world, along with Bangladesh. Of the 62 percent devoted to farmland, about 80 percent is used for growing fodder for farm animals.

4. Statistic Denmark, http://www.lf.dk/Tal_og_Analyser/Aarstatistikker/Statistik_svin/Statistik_svin_2013.aspx. The number of pig farms in Denmark has been decreasing. In 1993, there were 26,860 farms with pigs. In 2013, the number was 3,955. However, the number of, for instance, slaughter pigs (i.e., pigs that are actually slaughtered) is nearly the same for 1993 and 2013.

5. There are several strains of Methicillin-Resistant Staphylococcus aureus (MRSA) bacteria, most of which we carry, but one strain in particular, cc398, is associated with animals, and in Denmark it is found primarily in pigs. MRSA are staphylococcus bacte-

ria resistant to the antibiotic Methicillin (http://www.pigresearchcentre.dk/Pig%20 Production/MRSA.aspx).

6. Other specialized husbandry farm production in Denmark includes chickens, egg layers (i.e., hens kept specifically for egg production, as opposed to chickens raised for slaughter), and minks.

7. The final production of slaughter pigs has been reduced dramatically in recent years, because Danish abattoirs have cut back their capacity to slaughter, mainly because Danish wages are higher than those in other European countries. As a result, the weaners (pigs weighing between seven and thirty kilos) are exported alive to, for instance, Germany and Poland. A few months later, when they reach one hundred kilos, they are slaughtered there.

8. A few farms specialize in raising *polte* (young pigs). *Polte* is the Danish word for young sows that have not yet been mated. On these farms, they are produced as future sows for other farmers and for export. Furthermore, a few farms also specialize in producing semen from boars, as all pigs today are artificially inseminated, with semen selected for its genetic qualities.

9. No regulation requires pig farmers to straw. However, some producers use the deep-litter method with straw because this is required for the pig they produce, for instance, the so-called English pig, which is exported to the UK market. Most pigs live directly on the concrete slotted floor, which became popular in the 1970s. Access to rooting material can differ from farm to farm. According to the legislation, all animals in all types of facilities must have permanent access to sufficient rooting and enrichment material, but the list of what materials can be used is long. See "Self-Audit Scheme for Animal Welfare on Danish Pig Farms 2013": http://eng.vsp.lf.dk/~/media/Files/PDF%20-%20Viden/Branchekode_UK.ashx.

10. An average of thirty piglets per sow per year has been a success that has led to fierce public debates, because many piglets are born too small, dying just after birth or during their first days. Some twenty-five thousand newborn piglets die in Danish barns each day, and the ethical and welfare implication of large litter size in the domestic pig is debated in research (Rutherford et al. 2011).

11. Since January 2013, Danish national and EU legislation has required that pregnant sows be kept loose in groups from four weeks after service until one week before farrowing. On this farm they started taking pregnant sows into loose groups four years before the legislation became effective (European Commission 2011).

12. Fieldwork took place in connection with the research for Inger Anneberg's PhD (2013): "Actions of and Interactions between Authorities and Livestock Farmers—in Relation to Animal Welfare" (Anneberg 2013).

13. F. W. R. Brambell (chairman), Report of the Technical Committee to Enquire into the Welfare of Animals Kept under Intensive Livestock Husbandry Systems (London: Her Majesty's Stationery Office, 1965).

14. 1. Freedom from Hunger and Thirst—by ready access to fresh water and a diet to maintain full health and vigor.

2. Freedom from Discomfort—by providing an appropriate environment including shelter and a comfortable resting area.

3. Freedom from Pain, Injury or Disease—by prevention or rapid diagnosis and treatment.

4. Freedom to Express Normal Behavior—by providing sufficient space, proper facilities and company of the animal's own kind.

5. Freedom from Fear and Distress—by ensuring conditions and treatment which avoid mental suffering.

See https://www.gov.uk/government/groups/farm-animal-welfare-committee-fawc.

15. See http://www.globalmeatnews.com/Industry-Markets/EFSA-publishes-scientific-opinion-on-tail-docking-in-pigs.

16. The group included veterinarians, biologists, an agronomist, and an anthropologist.

17. http://ec.europa.eu/food/animal/welfare/farm/pigs_en.htm.

18. At a 2014 meeting in Copenhagen between the government and agricultural partners, it was suggested as a volunteer goal for the pig producers that 10 percent of the sows be loose in the farrowing pens by 2020.

REFERENCES

Anneberg, Inger. 2013. "Actions of and Interactions between Authorities and Livestock Farmers—in Relation to Animal Welfare." PhD thesis, Aarhus University.

Anneberg, I., and N. Bubandt. 2015. "The Animal Welfare State: The Body of the Pig, the History of Welfare and the Politics of Life Itself in Denmark." In Danish: DYREVELFÆRDSSTATEN. Grisens krop, velfærdens historie og selve livets politik i Danmark. Særnummer om velfærdsstaten. *Tidsskrift for Antropologi* 73 (2016).

Anneberg, I., M. Vaarst, and J. T. Sørensen. 2012. "The Experience of Animal Welfare Inspections as Perceived by Danish Livestock Farmers: A Qualitative Research Approach." *Livestock Science* 147: 49–58.

Anneberg, I., M. Vaarst, and N. Bubandt. 2013. "Pigs and Profits: Hybrids of Animals, Technology and Humans in Danish Industrialized Farming." *Thinking with Latour*, special issue, *Social Anthropology* 21, no. 4 (November): 542–59.

Callon, Michel. 1986. "Some Elements of a Sociology of Translation: Domestication of the Scallops and the Fishermen of St. Brieuc Bay." In *Power, Action and Belief: A New Sociology of Knowledge?* ed. John Law, 196–223. London: Routledge.

Cassidy, Rebecca. 2007. "Introduction: Domestication Reconsidered." In *Where the Wild Things Are Now: Domestication Reconsidered*, ed. Rebecca Cassidy and Molly Mullin, 1–25. Oxford: Berg.

Clutton-Brock, J. 1989. *The Walking Larder: Patterns of Domestication, Pastoralism and Predation.* London: Unwin Hyman.

European Commission. 2011. "Animal Health and Welfare: Pigs." http://ec.europa.eu/food/animal/welfare/farm/pigs_en.htm.

Forkman, B. 2011. Velfærdsvurdering og velfærdskontrol. Ministeriet for Fødevarer, Landbrug og Fiskeri, Fødevarestyrelsen.

Foucault, M. 1982. "The Subject and Power." *Critical Inquiry* 8, no. 4: 777–95.

Fraser, D. 2008a. *Understanding Animal Welfare: The Science in Its Cultural Context.* UFAW Animal Welfare Series. West Sussex: UFAW.
Fraser, D. 2008b. "Toward a Global Perspective on Farm Animal Welfare." *Applied Animal Behaviour Science* 113, no. 4: 330–39.
Fredeen, H. T., and P. Jonsson. 1957. "Genic Variance and Covariance in Danish Landrace Swine as Evaluated under a System of Individual Feeding of Progeny Test Group." *Zeitschrift für Tierzüchtung und Züchtungsbiologie* 70, no. 4: 348–63.
Gjerris, M. 2012. "The Willed Blindness of Humans: Animal Welfare and Beyond." In *Climate Change and Sustainable Development: Ethical Perspectives on Land Use and Food Production*, ed. Thomas Potthast and Simon Meisch. Wageningen: Wageningen Academic Publishers.
Gjerris, M. 2015. "Willed Blindness: A Discussion of Our Moral Shortcomings in Relation to Animals." *Journal of Agricultural Enviromental Ethics* 28: 517–32.
Hamann, K. 2006. "An Overview of Danish Pork Industry Integration and Structure." *Advances in Pork Production* 17: 93.
Harfeld, J. 2010. "The Question of the Farm Animal: Welfare, Ethics, and Public Policy in Modern Animal Agriculture." PhD thesis, Aarhus University.
Harrison, R. 1964. *Animal Machines: The New Factory Farming Industry.* London: Vincent Stuart Publishers.
Higgins, D. M., and M. Mordhorst. 2008. "Reputation and Export Performance: Danish Butter Exports and the British Market, c. 1880–c. 1914." *Business History* 50, no. 2: 185–204.
Hurn, S. 2012. *Human and Other Animals: Cross-Cultural Perspectives on Human-Animal Interactions.* London: Pluto.
Jenkins, R. 2012. *Being Danish: Paradoxes of Identity in Everyday Life.* Copenhagen: Museum Tusculanum Press, University of Copenhagen.
Johannsen, V. K., T. Nord-Larsen, T. Riis-Nielsen, K. Suadicani, and B. B. Jørgensen. 2013. Skove og plantager 2012, Skov & Landskab, Frederiksberg, 2013. 189 s. ill. http://ign.ku.dk/samarbejde-raadgivning/myndighedsbetjening/skovovervaagning/skove-og-plantager-2012.pdf.
Law, J., and M. E. Lien. 2014. "Animal Architextures." In *Objects and Materials: A Routledge Companion*, ed. Penny Harvey, Gillian Evans, Hannah Knox, Christine McLean, Elizabeth B. Silva, Nicholas Thoburn, and Kath Woodward. London: Routledge.
Lien, M. E. 2015. *Becoming Salmon.* Berkeley: University of California Press.
McGlone, J. J. 1993. "What Is Animal Welfare?" *Journal of Agricultural & Environmental Ethics* 6 (Suppl 2): 26–36.
Meyer, N., and K. Villadsen. 2013. *Power and Welfare: Understanding Citizens' Encounters with State Welfare.* Oxford: Routledge.
Natur- og Landbrugskommissionen. 2012. "Status Report—2012." http://www.naturoglandbrug.dk/statusrapport_2012.aspx?ID=51058.
Ploug, N. 2004. *Den danske velfærdsstats historie: En antologi.* Copenhagen: Socialforskningsinstituttet.
Politiko.dk. Netmedia. 2014. "Minister of Health Forbids Danish School Children to

Visit Pig Farms." In Danish: http://www.politiko.dk/nyheder/sundhedsminister-forbyder-besoeg-paa-danske-gaarde.

Rutherford, K. M. D., E. M. Baxter, B. Ask, P. Berg, R. B. D'Eath, S. Jarvis, K. K. Jensen, A. B. Lawrence, V. A. Moustsen, S. K. Robson, F. Thorup, S. P. Turner, and P. Sandøe. 2011. The Ethical and Welfare Implications of Large Litter Size in the Domestic Pig: Challenges and Solutions. Danish Centre for Bioethics and Risk Assessment (CeBRA). Project Report no. 17.

Scott, J. 2011. *The Tanner Lectures on Human Values: Four Domestications; Fire, Plants, Animals, and . . . Us.* http://tannerlectures.utah.edu/_documents/a-to-z/s/Scott_11.pdf: Harvard University.

Singleton, V. 2010. "Good Farming: Control or Care?" In *Care in Practice: On Tinkering in Clinics, Homes and Farms*, ed. Annemarie Mol, Ingunn Moser, and Jeannette Pols. Bielefeld: Trancript Verlag.

Sorensen, J. T., and D. Fraser. 2010. "On-Farm Welfare Assessment for Regulatory Purposes: Issues and Possible Solutions." *Livestock Science* 131, no. 1: 1–7.

Tuyttens, F. A. M., F. Vanhonacker, B. Verhille, D. De Brabander, and W. Verbeke. 2012. "Pig Producer Attitude towards Surgical Castration of Piglets without Anaesthesia versus Alternative Strategies." *Research in Veterinary Science* 92, no. 3: 524–30.

Veissier, I., A. Butterworth, B. Bock, and E. Roe. 2008. "European Approaches to Ensure Good Animal Welfare." In *Applied Animal Behaviour Science* 113: 279–97.

Wemelsfelder, F., and S. Mullan. 2014. "Applying Ethological and Health Indicators to Practical Animal Welfare Assessment." In *Animal Welfare: Focusing on the Future*, ed. D. J. Mellor. OIE *Scientific and Technical Review* 33, no. 1.

5

DUCKS INTO HOUSES

Domestication and Its Margins · *Marianne Elisabeth Lien*

One of the most problematic companions to the idea of domestication is that of civilization. As detailed in the introduction to this volume, the narrative of the Neolithic Revolution marked that watershed moment in which domestication got embedded in time, performing a distinction between the domesticated and the wild, the civilized and the savage. But its accuracy as a classifier of stages of human progress is hugely exaggerated: while there is no doubt that the cultivation of grain in the Middle East several thousand years ago significantly altered the conditions of life for many people (and not only for the better), it is also true that for many other *people living elsewhere*, cultivation of grain did not make much of a difference or was never an alternative.

Such is the situation, for instance, for people living in the subarctic. In relation to *their* practices and predicaments, the significance of domestication is not about agrarian cultivation as such. Instead, domestication has come to matter as an ordering device, sorting landscape practices and people along a temporal trajectory of progress according to modes of subsistence. This distinction has inscribed boundaries in the landscape, between infields and outfields, between property and commons, between migratory and sedentary livelihoods, and between use and conservation—boundaries that have sometimes rendered indigenous and local landscape practices invisible, illegal, or severely restricted in scope.[1] In this way, the conceptual twinning of domestication and civilization has highlighted specific modes of cultivation (owning, plowing, fertilizing) as signs of human progress, rendering other food procuring practices marginal, primitive, or obsolete. For people in the subarctic, then, the narrative of domestication as a story of human progress has served as a tool of oppression, a colonializing device in its own right.

Domestication at the Margins: Domestication in the North

This chapter explores domestication in the Scandinavian subarctic, or the Circumpolar North. Often described (treasured or discarded) as barren and remote, northern places have typically been portrayed as a wilderness to be conquered rather than a landscape to inhabit (Bravo and Sörlin 2002). This makes the subarctic particularly relevant for a reconsideration of domestication. First, the region has systematically been marginalized by nation-building narratives and state policies that see agriculture as a sign of progress or development. Such marginalization of the subarctic livelihoods can be seen as the flipside of the enhanced academic interest in the fertile regions farther south, which were also sites of early Neolithic cultivation: The idea of the fertile crescent as the cradle of "civilization" was initially presented precisely as a contrast to "barbarian" regions further north. Consider, for example, the very first sentence of Gordon's Childe's famous book *The Most Ancient East* (1928, 1):[2] "Barely a thousand years ago Scotland and the rest of northern Europe were still sunk in the night of illiteracy and barbarism." The rest of the book highlights the Ancient East, which represents "an indispensable prelude to the true appreciation of European prehistory." We see how civilization is performed through othering the so-called noncivilized, and how acknowledging the historical significance of cultivation of plants and husbandry animals involved a simultaneous marginalization of those lands where animals and plants were engaged with differently, such as regions farther north.

Second, human-animal relations in the Arctic defy many of the common assumptions associated with a conventional notion of domestication as a practice of domination, control, or confinement. Northern landscapes are indeed cultivated places. Even when they are not cultivated by "plow and spade," human engagements with Arctic landscapes comprise a rich array of practices that involve listening, watching, participating, and shaping (see, e.g., Anderson 2004:2). Such practices require a complex understanding of the movements and habits of hunted, fished, or herded animals, and this understanding is often anchored in concepts that exceed biological taxonomies (Ween 2012, Joks and Law 2017; Rybråten 2014). Research in Arctic regions further suggests that mobility, fluidity, and adaptability to shifting seasonal conditions and animals' migrations take on greater importance (Kramvig 2005; Ween and Lien 2012). Hence, sites that appear barren and wild to those who are accustomed to seeing planted fields are not "wilderness" at all but have been shaped by practices of care and cultivation. Consider, for example, how Sámi fishing practices involve active caring for rivers and lakes (for details, see Law and Østmo 2016). Local practitioners resist calling this domestication but describe the practices as ongoing relations and alterations of river and lake habitats that, over time and in rather subtle ways, can shape conditions for the species involved. Such relations are worth studying in their own right, as they can broaden our understanding of the scope of human-animal coexistences in landscapes where the human "footprint" is hardly visible.

Third, as subarctic regions include landscapes where the plants that "cultivated continental Europe" simply cannot grow, these regions are often zones of colonial conquest. At the edge of viability for common agricultural assemblages, they become experimental sites of development, subject to state efforts to "domesticate" its national peripheries, or sites for the extraction of untapped resources. Such state projects have shaped landscapes and human predicaments in ways that have often undermined the resources and landscape affordances that local inhabitants rely upon. Understanding more-than-human relationships in the subarctic has become more urgent as the region has become the focus of extractive and industrial expansion and as regulations and restrictions have been placed on landscapes through nature conservation and commercial recreational use.

Exploring domestication from such margins can yield important insight into practices and relations that are otherwise overlooked. It can also yield insight into the processes whereby such relations become disrupted, marginalized, or silenced and thus illuminate how dominant tropes associated with domestication are linked to colonizing processes that potentially threaten the maintenance of northern livelihoods (Lien 2014).

Northern Livelihoods and Fishermen-Farmers

My regional focus in this chapter is the North-Norwegian coast, the rugged coastline that stretches from just below the Arctic Circle, curves northeast, and then bends south to where it meets the Russian border, encompassing windswept islands, inlets, and rich fishing grounds. Humans have inhabited this fertile edge between land and ocean for nearly ten thousand years, hunting, fishing, trapping, herding, and engaging in varied forms of gathering (Hansen and Olson 2004). Gradually, these practices were supplemented by animal husbandry (sheep, goats, milking cows) and by the extended cultivation of small plots of grassland for animal fodder and grazing. *Fiskarbonden*—the fisherman-farmer—is a Norwegian term commonly used to describe the specificity of such coastal households in the north (Brox 1966). A fisherman-farmer is typically a household that relies on a broad variety of food-procuring practices, including fishing and small-scale subsistence farming, augmented with a number of other activities such as gathering bird eggs, picking berries, hunting, trapping, and, in certain areas, collecting eiderdown. Traditionally, such practices often involved considerable seasonal migration, too. These subsistence modes were the norm until the mid-twentieth century and in many places much longer. Thus, the term *fiskarbonden* reminds us that Northern Norwegian farming was different from the very beginning and also marked as such in a discourse that was modeled on the agricultural practices of most of Europe, as well as Southern Scandinavia.[3]

Although fishing and trade in fish constituted the backbone of the coastal economy in the north, people's livelihoods have always relied on a range of activities that were seasonal and intertwined with extensive knowledge of the local landscape and seascapes. The fisherman-farmer's reliance on a broad set of food-procuring practices reflects the limited economic value of "the farm" itself: close to the Arctic Circle, most plants that cultivated Europe cannot be sustained. Hence, viability in this region is not so much about ploughing the field as about knowing how to make use of the affordances that the landscape and seascape can provide, and not least to be in the right place at the right time. This is expressed in the Sámi term *meahcci*, a concept that is related to place, movement, seasons, and affordance in the landscape. Norwegian vernacular maps these landscapes as *utmark* (literally outfields), in contrast to *innmark* (infields), which denotes the gardens and fields of the sedentary farmer (Ween and Lien 2012). Although the Norwegian term *utmark* is often (mis) translated to the Sámi term *meahcci*, the notion of *meahcci* is broader and more dynamic, as it bears witness to the entanglements of persons, animals, stories, and plants that together temporarily constitute the landscape and seascape as valuable (see Rybråten 2014, Ween 2017). Hence, they provide an important contrast and a situated backdrop to other and overlapping state practices that mark farmland as a resource—practices associated with the Norwegian (formerly Danish Norwegian) nation-state.

Arctic and subarctic regions provide insight into human-animal relations that do not fall easily into the conventional models of domestication discussed above (see also the introduction). Charles Stépanoff (2012), for example, describes reindeer-herding relations in Southern Siberia as an ongoing cooperative context in which herders rely on the animals' autonomy, cognitive skills, and desires to engage the herd. In a similar vein, Anderson (2014) maintains that herding units in the Arctic are often associated with what he calls "cultures of reciprocity," in contrast with "cultures of control," which are often attributed to relations of domestication (see also Paine 1988; Bjørklund 2004). Joks's and Law's study of Sámi fishing practices in the river Deatnu similarly suggest a relation in which the fish, rather than passive prey, enters a relation of mutual responsiveness. This is captured in the Sámi term *bivdit*, which, on the one hand, indexes fishing and, on the other hand, means to plead or to ask for: "Fishing then is partly about asking, respectfully. The salmon may accept the request, or it may not. In a certain way it has autonomy" (Joks and Law 2017, 5; see also Joks 2015). Similarly open-ended and mutual relations unfold along the Northern Norwegian coast. On the coast of Helgeland in Nordland County, and especially around the Vega Archipelago, the relations between humans and birds have taken a peculiar turn. Here, eider ducks (the common eider) have sought nesting sites in close proximity to human settlements, while humans have built nesting houses for easy access to the eggs and eiderdown. The eiderdown has been an important additional source of income for the fishermen-farmers and has been sold as a highly priced luxury product to European nobility since the 1500s (Fageraas 2016).

Neither domesticated nor wild, the eider ducks on the Vega Islands alert us to how subtle and repeated encounters between humans and animals have come to shape more-than-human worlds in mutual processes of biosocial becoming (Ingold and Pálsson 2013). These relations are always uncertain, open-ended, and potentially transformative and hence might inspire us to think domestication differently. Guided by ethnography of the eider ducks, this chapter aims to recapture domestication as a reflexive tool and as an orientational point of departure for ethnographic inquiry (and curiosity) in relation to more-than-human encounters. Instead of aiming to classify the bird as either domestic or wild, I suggest that we mobilize domestication as a signpost to direct our gaze—perhaps to slow it down—in a way that allows us to notice the dynamic unfoldings of interspecies relational practices, wherever they may be. The point of doing so is not to extend the term "domestication" to every instance of cross-species interaction but rather to draw attention to the rich and complex worlds of interspecies relations that human engagements with other species can involve and that we may fail to grasp if we assume that we know precisely what domestication entails. This move is also political: drawing attention to the subtle practices of domestication that tend to escape from view (for academics, conservationists, and state bureaucrats alike), we are also better able to identify the policies through which such relations are at risk of being undermined or made absent. But first, let us briefly revisit the notions of domestication as destiny and domestication as control.

Domestication beyond Control

Domestication is more than confinement and control. Consider, for example, the Seejiq in Taiwan, an indigenous group who were supposedly among the first Austronesians to domesticate fowl, pigs, and dogs. According to anthropologist Scott Simon, Seejiq people nurture relations with these animals that are hardly captured by conventional definitions of domestication as control: "They live with dogs as hunting partners, in relationships of trust and friendship, rather than as creatures to dominate. The Seejiq can barely be said to have domesticated pigs, which are important for sacrifice but best kept at a distance" (Simon 2015, 706). Contributions in animal studies also indicate that even from a situation of relative confinement, nonhumans act upon us in ways that are unexpected and defy human attempts at control (see Schroer, this volume). Even in agro-industrial settings, such as salmon farms (Law and Lien 2013; Lien 2015), or industrial pig stalls (Anneberg and Varst, this volume) there is a biosocial liveliness that runs the risk of being rendered flat by descriptions of domestication that take tales of human mastery at face value.

Partly, these insights reflect an approach that admits, or is attentive to, the possibility that animals may exert some kind of agency, for example, through fear, hunger, or desire. In this way, the animals, fish, or birds may contribute to shaping the human-animal relation in a process that has been described as co-domestic (Fijn 2011). Such insights have also informed the interpretation of archeological material and the theories of how human-animal relations affected processes of (natural) selection under domestication. Many archeologists argue, for example, that dog domestication was initiated by wolves who began to specialize in feeding within human areas on the remains of leftover meals, particularly hunted animals. The assumption is that wolves gradually evolved (through selective pressures) to have less fear of humans and that human intervention in their breeding occurred long after this initial colonization of the human niche by the animals themselves (Coppinger and Coppinger 2001; Zeder 2012). In a similar vein, consider the puzzle of why guinea pigs were "chosen" as house companions in South America when so many other endemic American mammals seem more attractive as potential domesticates. According to archaeologist O'Connor (1997), it was the guinea pigs themselves that started the association by scavenging around human settlements (O'Connor 1997, 155). Humans, in turn, accommodated this move, allowing and nurturing an ongoing relation. This brings our attention to the intimate and reciprocal set of interactions that thus unfolds between humans and animals. Pat Shipman (2010) has suggested that a universal trait that sets humans apart from most other animals is what she calls an "animal connection." In this perspective, domestication is not about control but fundamentally about developing a means of communicating with other species, across difference. Such perspectives are useful when considering human-animal relations in the Arctic.

Yet the notion of domestication as control, and domestication as destiny, continues to haunt.[4] Poised between "the wild" and "the domestic," reindeer are routinely described as "*semi*-domesticated."[5] Similarly, human-animal relations that do not map easily onto the nature-culture dichotomy that the notion of domestication performs are rendered "incipient," or "not quite domestic" (cf. Vigne 2011).[6] Haunted by the distinction between wild and domestic, human-animal relations in the Arctic risk either being overlooked or being classified in relation to a historical/temporal trajectory that is inappropriate.

A similar tendency is found in historical and popular accounts, such as when Australian Aboriginals' "failure" to domesticate animals is explained with reference to the absence of large mammals on the Australian continent (as if certain preconditions necessarily take people down the lane of domestication), instead of seeing Aboriginal ways of "caring for country" as distinctive and complete practices in their own right (see Fijn this volume).[7] Titles like "The Biggest Estate on Earth" take a different view, as historian Bill Gammage alerts his readers to the details of Aboriginal so-called land-management strategies on the Australian continent that transformed the land before the arrival of Europeans (Gammage 2011). The intervention is important, but why are Aboriginal Australians portrayed as incipient gardeners? Why "management" and why "estate"? What is lost when diverse and complex forms of landscape practices are portrayed through the models of European farming?

Rather than approaching Arctic human-animal relations as hybrid instances, or as domestication not "fully accomplished," this chapter suggests that such regions may instead radically *broaden* the notion both of what domestication might be and of the multispecies entanglements that it engenders. Furthermore, they shift the scene of domestication from one-to-one relationships between people and single species to landscape configurations (Anderson 2000b; Tsing 2012; see also Swanson this volume). Incorporating *landscapes* in domestication accounts helps us replace narratives about confinement and control with approaches that examine the complexities of changing multispecies relations in which humans play an important role but are not "in charge." Let us return to the eider ducks at the Vega Archipelago.

Ducks into Houses

The first thing you might notice as you arrive by boat to one of the many islands in the Vega Archipelago is a wooden cross carrying an inscription "FREDLYST" (see figure 5.1). *Fredlyst* is an archaic proclamation that literally translates as peace. Sticking a wooden cross into the thin layer of soil on these remote Norwegian islands was an ancient way of announcing the presence of nesting eider ducks.[8] The cross signaled that somebody cared for the birds and that other people ought to stay away, or, if they needed to go onshore, that they must be very careful and quiet. The ancient crosses are long gone, but as the eider duck practice has been revived in recent years, crosses have emerged too.

Eider ducks have been cared for on the islands at least since Viking times, possibly much longer, and the seabirds have been hunted and harvested for as long as the coastline has been inhabited (Berglund 2009). During the past millennium, a specialized production has evolved in which locals build nests and "houses" for the birds and collect eiderdown and eggs in return. The seasonal practice of attending to the rookeries and the birds constitutes one of the many landscape practices that were required to sustain oneself and one's family on these islands. In what follows, I draw on Norwegian anthropologist Bente Sundsvold's ethnography (herself native to Vega) and her detailed account of a peculiar practice that state legislation and ignorance nearly made extinct but that has recently been reestablished, due in part to international recognition associated with the islands designation as a UNESCO World Heritage site in 2004.

The annual encounters between ducks and their people take place in spring, when the eider (*Somateria mollissima*), locally referred to as *éa*, come onshore to nest. With thousands of islands to choose from (most of them uninhabited

FIGURE 5.1. The ancient *fredlyst* crosses are long gone, but the practice has been revived in recent years. This cross has been erected by Aud Halmøy. Photo by and with permission of Bente Sundsvold.

by people), they could go anywhere, but over the course of many generations, nesting ducks have found their way to where people are. Hence, during the brooding season, which lasts about a month, "people live in the midst of a colony of nesting eiders" (Sundsvold 2010, 2015). Carefully prepared, with roofs on top and dried seaweed inside, the nest boxes offer a site of protection against rough weather as well as predators (figure 5.2).

Some nesting sites are placed against the wall of a house or within a cluster of buildings. Others are assembled on otherwise uninhabited islands and islets. Sundsvold's collaborator, Aud Halmøy, one of the pioneers of revitalizing eider duck practice, prepared more than 150 nests in one season, about three times as many, she reckons, as the number of nests that will actually be used. This way, the ducks will get to choose the nests they find most appropriate. Assembled from driftwood, leftover planks, slabs of stone, or simply an old boat turned upside down, such makeshift "domuses" are visual reminders of human presence in this remote island landscape.

Locally, as Sundsvold vividly describes, the brooding season had a special atmosphere and a special name: *varntie*. The prefix *varn* may be translated to "care," "precaution," or "protection," while *tie* refers to time (Sundsvold 2010, 96), but it is a word that few Norwegian speakers today would understand. As

FIGURE 5.2. Eider duck houses. Photo by and with permission of Aud Halmøy.

a "linguistic ghost" from another era (Mathews 2017), it enacts a landscape in which relations were done differently, when human activities were restricted by the presence of birds. *Varntie* was the time when everyone kept quiet, when dogs and cats were on leash or inside and when walking was restricted to the old established footpaths to which the birds were accustomed (Sundsvold 2010). Smoke from fireplaces was avoided, if possible (Fageraas 2016). Children were told to move slowly, speak softly, and pay attention. If they encountered a bird's nest, they were to gently step aside. Today, similar precautions are practiced by community residents. If you move carefully, or "think like a bird," as Aud Halmøy put it, they will slowly come to accept your presence. Others recall that those who cared for birds did not wash their clothes, in order to make it easier for the birds to become familiar with their people (Fageraas 2016). Sundsvold describes the process of caring for birds as a gradual earning of the birds' trust. The clue, she suggests, is to make oneself predictable to the birds. Achieving that requires an effort from everyone present, and the term *varntie* is a way of capturing this exceptional time. This is a term that cuts across distinctions between humans and their natural surroundings and instead fixes a shared temporal slot that is set apart for humans and birds alike.

But what attracts the ducks? For a human observer, the most obvious at-

FIGURE 5.3. Nesting eider duck. Photo by and with permission of Bente Sundsvold.

tractions are the nesting sites themselves. With a solid roof and dry seaweed bedding, they offer a ready-made comfort that nesting ducks can obviously appreciate (figure 5.3).

Another key amenity is the protection against predators that this arrangement offers. The roofed houses make the nesting sites less visible to otters and eagles. But the rookery is only a partial protection, delaying the attack or diverting the predators' attention to nesting sites in the open. It offers no guarantee. This is where human presence becomes significant. Their mere presence can scare the predators away, and active human intervention can make a difference, too: now and then, the silence of *varntie* is shattered by a gunshot in the air. Aud Halmøy always brings a gun with her on the daily rounds. Occasionally a crow or a black-backed gull is shot. And before the season starts, mink traps are distributed to avoid the devastating effects of a recent predator (see below).

Are the eiders tame? Bente Sundsvold insists that they are not. This, she argues, is why the *varntie* was so important. The idea is that a successful brood of ducklings will cause the female eider to return year after year. Later, as they mature, the ducklings will return, too, securing additional income. When Aud Halmøy lost nearly the entire stock of eiders due to a particularly nasty predator attack at the rookeries she had prepared on an uninhabited islet, she was devastated. Certain that years of effort to earn the trust of the ducks was lost, she was almost ready to give up. But much to her surprise, the eider ducks returned in large numbers the following year, though their preferences had changed: instead of nestling on the slightly remote islets, they now chose to nest near the cluster of houses. They chose the rookeries that would be much less quiet but that would offer more protection against predators, due to the continuous presence of Aud and her family. Perhaps, then, the predators too are a necessary part of this domestication assemblage? Sundsvold claims that they are: without predators, she argues, it is unlikely that the birds would have sought these rookeries in the first place. In this way, the presence of otters and eagles facilitates the human gathering of eiderdown and eggs. Another key component is some form of memory across bird lives and generations: As is the case for salmon (see Ween and Swanson, this volume), the relation of domestication hinges on migratory returns of the same birds, year after year, and then, later, of their offspring. To build up a *dunvær* takes many, many years; it is a gradual institutionalization of trust, articulated through the brooding *éa*'s selection of house, and passed down through generations of birds, as well as people.[9]

Domestication Reconsidered

Seen through a conventional approach to domestication, these eider duck arrangements are barely visible. Like many other instances of multispecies relations in the North, the story of the eider ducks challenges our categories of wild and domestic. Neither confined nor controlled—neither owned nor fed—the ducks have none of the characteristics that place them in the category of conventional domesticates (cf. Clutton-Brock 1994). And yet, the assemblages have been consequential for birds as well as for people. According to biological research on eider ducks on Spitsbergen, it is quite likely that human-made protected shelters have had an impact on the duck populations' survival rates. If so, it might also historically have changed selection pressures, possibly favoring those birds that tolerated human presence. For humans, it has provided food as well as income and thus contributed to a more robust livelihood in a coastal landscape where fish were abundant but farming was precarious, pasture was scarce, and grain-cultivation nonexistent.

Seen in relation to more recent debates on domestication, however, the duck and down assemblages at the Vega islands are a highly relevant case to consider, and one that can teach us about a kind of variation in domestication practices that is often overlooked. We see the unfolding of a social relation that is both asymmetrical *and* simultaneously built on trust. We see how presumably wild birds seek human compounds for protection, like so many animals have done before them. Finally, we see how the relations that unfold involve a complex interspecies relation of ducks-predators-humans, a relation that challenges the assumption of domestication as a binary process. Perhaps the relation is best described as what Natasha Fijn, with reference to Mongolian herding practices, calls a codomestic relationship: "the social adaptation of animals in association with human beings by the means of *mutual* cross-species interaction and social engagement" (Fijn 2011, 19, emphasis in original). According to Fijn, relations of codomestication are spatially situated, encompassed within the landscape, or codomestic sphere, which in the case of the Mongolian herders is synonymous with their encampment (Fijn 2011, 220).

According to historical sources, humans have gathered eiderdown in Vega for several hundred years, possibly for more than a millennium. Does this qualify as an incipient phase in a "gradual process of domestication"? Is it a phase on a journey from wild to domestic, from trust to domination, or from being "free as a bird" to being subject to tightened regimes of control and confinement? Based on Sundsvold's and others' accounts, we may reasonably conclude that it is not. That "ducks go into houses" does not mean that they have come to stay, nor that they are, in evolutionary terms, on a trajectory toward a final destination (cf. Zeder 2012, 91). Their makeshift domus is a temporary amenity, a material expression of a fragile relation of codomestication involving birds and people that offers no guarantees for anyone and that can easily fall apart. For Mongolian herders, these relations are spatially situated within their encampment. In Vega, the relation is also encompassed temporarily, in the specific time known as *varntie*. In this way, the *dunvær* assemblages remind us that domestication is—perhaps always—a precarious achievement, a sensibility that is easily lost when our assumptions about domestication are too committed to notions of confinement and control. They show that we should not let hegemonic idioms of European farming narrow our gaze.

This is important everywhere but especially at northern sites like Vega, where conventional farming is hardly enough to feed a family. In such regions, livelihoods have typically relied on a broad range of landscape practices and on subtle of ways knowing the land and the animals in their surroundings.[10] That some such practices were institutionalized is therefore not surprising. As Sundsvold explains, *fredlysning* means to publicly announce the consecration of peace, "in the medieval sense of a public proclamation" (Sundsvold 2010).[11] She describes how *fredlysning* sites have been registered in the Vega islands, since the eighteenth century, but that the practice was probably much more common than these legal registrations indicate.[12] *Fredlysning* was not a proclamation of ownership in the sense of announcing private property but rather a temporary institutionalization of specific user rights and user interests that regulated the relations between people and environmental resources in a region where farming and private property were of limited importance. The institution was still practiced in Vega in the 1950s and 1960s. Most of the proclamations of *fredlysning* involve eider ducks, so-called *dunvær*, but some concern cloudberries and other amenities, such as specific fish used for bait, grass that can be cut and used as fodder, and in the old days even otter, which was hunted for fur. As *fredlysning* did not require ownership of land, it served to secure a certain income or access to resources for people who were otherwise disenfranchised or had limited sources of livelihood.

During the twentieth century, the Norwegian state had made its presence more strongly felt. Especially from the 1950s onward, state efforts at centralization were commonplace and rather efficient. Anthropologist Ottar Brox voiced a solid critique of state policy in Northern Norway during the postwar era. His key argument is that through various subsidy regimes that sought to enhance capital growth by favoring those who were specialized as farmers only, or fishermen only, the state ignored the uniquely adaptive subsistence strategies of fishermen-farmers of the Northern Norwegian coast (Brox 1966, 66). This made traditional livelihoods increasingly difficult to sustain. The push toward urbanization was further exacerbated by the state's classification of coastal hamlets as either places one should move to (*tilflyttingssted*) or places one ought to move away from (*fraflyttingsted*), accompanied by state subsidies for those who decided to move (Sundsvold 2010).[13] The largest village on the central island in the Vega Archipelago was a place to move *to*, while all other settlements on the remote islands were places to move *from*. In the same period, electrification, concentration of schools, and the acquisitions of motorized fishing vessels all contributed to making the more remote islands less attractive. In spite of such measures, many remained, and the main exodus from the islands to the new center did not happen until the 1970s—not until the arrival of mink farming.

By the 1970s, state-subsidized mink farming had been promoted by regional developers, and a number of farms had been established, even in the Vega islands. Mink (*mustela vison*) were introduced to Norway from North America in the 1930s, and the first farms were established in the coastal region of Helgeland in the 1940s. Soon, a feral population of mink was established, too, and according to Sundsvold, its impact on the domesticated ducks was catastrophic, as feral minks became a new and efficient predator. Why, she asks, did the islanders not resist the establishment of mink farms side by side with eider duck houses? Why was it so easy to place wild animals in cages (118)?

The answer, she suggests, has to do with notions of progress and the calls for rural economic development that were the mantra of the postwar era. Mink farming was potentially profitable, mobile, and scalable and could provide remote communities with the "development" that they sorely needed. I would add that the fact that mink were confined, controlled, and provided fur that could fetch a good price on the global market made it an obvious commodity for the imagined postwar future. That it simultaneously undermined another ancient livelihood, another precarious form of human-animal domestication, was not enough to keep it out.

Did the state completely ignore the presence of the ducks on the islands and their importance for the islanders? A final historical shift indicates that this might have been the case. In 1981, after having been inscribed in legal regulations for 140 years (and informally practiced much longer), the ancient institution of *fredlysning* was suddenly abolished. The occasion was a new set of regulations for hunting game.[14] If the introduction of the mink was catastrophic for the traditional duck and down rookery assemblage, this was, according to Sundsvold, the final nail in the coffin. With the implementation of the new regulations, the few remaining nesting sites that were still maintained could no longer protect the birds during brooding season. From now on, the landscape was open for all citizens on equal terms, as state-owned territory tends to be in Norway, where the right to roam is an important principle securing access.[15] The change signals a recognition of Norwegian landscapes as fundamentally state-owned and equally accessible to all, as well as an emergent recognition of landscapes as recreational wilderness, known in Norway simply as *natur* (nature).[16] As *natur*, the Vega islands and all its inhabitants are no longer exclusively known and cared for by locals but by the state Ministry of Environment, protected through policies of nature conservation and known through natural science. This is in contrast to agricultural land, husbandry animals, and pasture, which are regulated by the Ministry of Agriculture. In this way, human-animal relations that were not recognized as husbandry, but that were also not "not-husbandry," are matter-out-of-place, ignored or marginalized. Similarly, locally recognized connections between certain kin groups and certain cloudberry marshes that were once proclaimed through *fredlysning*—another subarctic practice of food procurement that does not fit the model of modern agriculture—are also ignored. In other words, as the Vega islands have become part of a state-managed Nature, they have simultaneously been alienated from the practices that once shaped people's livelihoods, practices that do not operate according to the binary of Nature and Culture.[17] At odds are two landscapes and two sets of human-animal relations: one is fluid, precarious, fragile, and largely invisible for state developers; the other is defined, protected, highly visible, and classified according to the hegemonic binary of the wild (which can be harvested, hunted, and used as a site for recreational pleasure) and the domestic (which belongs in cages). The brooding ducks' attentive clumsy steps toward makeshift houses are matter-out-of-place, prey to the feral mink, and invisible to the new hunting regimes. The fragile relations of codomestication that made the islands a codomestic sphere are made absent through the state regulation of outdoor life through equal right-to-roam for all.

Conclusion

Conventional approaches to domestication fail to appreciate the rich plethora of relational practices that are largely unpredictable and beyond control. A narrow approach to domestication, emphasizing control and confinement, can be a strong foundation for cultural critique (see, for example, Tsing, this volume). At the same time, the danger of such a narrow approach is that it marginalizes relational practices that do not fit the model. Often, and for reasons I have sketched above, such human-animal assemblages are already under threat, or rather precarious. In such places, marginalizations truly matter.

It is no coincidence that the site I chose to mobilize for this discussion is a cluster of islands near the Arctic circle. This is where state-subsidized, scalable development projects and ancient local landscape practices collide, or are practiced side by side, in awkward entanglements. "Ducks into houses" signal a literal movement of birds into makeshift houses, but it also gestures toward state efforts to encompass and to fix (perhaps even domesticate) the unruly complexities of island livelihoods through various political and legal reforms in the postwar era. The story of the eider ducks at Vega might be a case where, in the words of Marisol de la Cadena, "heterogeneous worlds that did not make themselves through the division between humans and nonhumans [are] *both* obliged into that distinction *and* exceed it" (de la Cadena 2014, 253, emphasis in original).

Proposing domestication as a reflexive tool is an attempt to make such heterogeneous worlds visible. Dismantling the tight association between the term "domestication" and notions of control and confinement compels us to take seriously the many different practices that do not fit the master narrative and that are consequently made absent in state management and legislation. Mobilized as a tool of orientation rather than as a classifying device, domestication can serve as a strategy for discerning the diverse practices that unfold in more-than-human relations and that defy, rather than depend upon, human mastery and control.

Acknowledgments

I gratefully acknowledge the support of the Norwegian Centre for Advanced Study (CAS), which funded our research project "Arctic Domestication in the Era of the Anthropocene," during the academic year 2015–16. This chapter has benefitted from generous advice from Diane Gifford-Gonzalez, Natasha Fijn, and Rob Losey, as well as my coeditors, Heather Anne Swanson and Gro B. Ween. I am particularly indebted to Bente Sundsvold for bringing to my attention eider ducks and for her detailed ethnography and thoughtful comments, and to CAS fellows for their discussions and inspiration.

NOTES

1. See, e.g., Beach (2004) and Sara (2009). This situation is not unique to the Arctic. See Nustad (2015) and Nustad, this volume, for a similar situation in South Africa.

2. *The Most Ancient East* is the book in which he first coined the term "revolution" in connection with the shift to Neolithic cultivation.

3. The fisherman-farmer denotes a gendered division of labor, in which men were at sea while women were the farmers (Brox 1966). Though they were hardworking, the women in these regions have traditionally held a relatively strong position and partly escaped the gendered subordination often associated with agricultural households (see Tsing, this volume).

4. The issue of control is also elaborated in a seminal paper by Tim Ingold, who asserted that the change in human-animal relations from hunting to pastoralism is fundamentally "a transition *from trust to domination*" (Ingold 2000, 75, emphasis added). While the hunter's tools are caught up in chains of personal causation and involve knowing the animal in a nuanced manner, the instruments of herding, he claims, are those of control: "They include the whip, spur, harness and hobble, all of them designed either to restrict or to induce movement through the infliction of physical force, and sometimes acute pain" (Ingold 2000, 73). But what about the sawbuck pack saddle, skillfully designed to distribute the weight evenly on the horse's back, the metal shoe protecting its hooves, the padding placed underneath the blanket, and the straw spread on the floor in the barn? Clearly, the instruments of domestication are many, and they are not all about control.

5. Nieminen (2011), http://www.miljodirektoratet.no/Global/dokumenter/horinger/Regelverk/Nieminen%20mfl.%202011.pdf?epslanguage=no.

6. See Anderson (2014) for a related critique.

7. I am grateful to Natasha Fijn for pointing this out.

8. The Vega Archipelago is a cluster of approximately 6,500 islands off the coast of Nordland County, just south of the Arctic Circle. Since 2004, it has been listed as a UNESCO World Heritage site. http://whc.unesco.org/en/list/1143.

9. *Dunvær* is the local name for the duck-rookery-down assemblage. *Dun* means down (feather), and *vær* is a common name for sites associated with particular affordances, such as *fiskevær* (where fish is landed).

10. This analysis focuses on coastal Northern Norway, but similar arguments could be made for other subarctic coastal regions, including Iceland, Scotland, the Faroes, and parts of Russia.

11. In old Norse language, *frid (fred)* connotes not only the absence of strife but also a relation of honor among free men.

12. *Fredlysning* is mentioned several places in legal documents from the Viking era, such as in the *Frostatingsloven*, which is known to exist at least since the year 1000 and which was written down in 1200.

13. The subsidies began in 1950 and remained for the next thirty-seven years, dismantled just fifteen years before the application to grant the Vega islands World Heritage site status (Sundsvold 2015).

14. *"Lov om jakt og fangst av vilt, 28. mai 1981."*

15. This is anchored in the legal term *allemannsrett* (all-man's-right).

16. This right-to-roam had already been legally established in 1958 with the so-called *Friluftsloven* (outdoor life legislation). See https://lovdata.no/dokument/NL/lov/1957-06-28-16. The new hunting regulations confirmed the principle of general access for hunting and fishing, as well.

17. This situation bears resemblances to the cultivated forests in Northern Italy. According to Andrew Mathews (2017), G146: "Over the last one hundred and fifty years, industrialization, rural out-migration, the arrival of alternative forms of fertilizer, and the arrival of successive epidemic diseases have undermined chestnut cultivation and transhumance, leaving a ruined landscape that is haunted by material and linguistic ghosts."

REFERENCES

Anderson, David G. 2000. *Identity and Ecology in Arctic Siberia: The Number One Reindeer Brigade.* Oxford: Oxford University Press.

Anderson, David G. 2014. "Cultures of Reciprocity and Cultures of Control in the Circumpolar North. " *Journal of Northern Studies* 8, no. 2: 11–27.

Anderson, David G., and Mark Nuttall. 2004. *Cultivating Arctic Landscapes.* Oxford: Berghahn.

Anderson, Virginia de John. 2004. *Creatures of Empire: How Domestic Animals Transformed Early America.* Oxford: Oxford University Press.

Barad, Karen. 2003. "Posthumanist Performativity: Towards an Understanding of How Matter Comes to Matter." *Signs: Journal of Women in Culture and Society* 28, no. 3: 801.

Barrett, Ronald, Christopher W. Kuzawa, Thomas McDade, and George J. Armelagos. 1998. "Emerging and Re-emerging Infectious Diseases: The Third Epidemiologic Transition." *Annula Review of Anthropology* 27: 247–71.

Beach, Hugh. 2004. "Political Ecology in Swedish Saamiland." In *Cultivating Arctic Landscapes*, ed. David G. Anderson, 110–24. Oxford: Berghahn.

Berglund, Birgitta. 2009. "Fugela Federum in Archeological Perspective: Eider Down as a Trade Commodity in Prehistoric Northern Europe." *Acta Borealia* 26, no. 2: 119–35.

Bökönyi, Sandor. 1989. "Definitions of Animal Domestication." In *The Walking Larder: Patterns of Domestication, Pastoralism and Predation*, ed. J. Clutton-Brock, 22–27. London: Routledge.

Bravo, Michael, and Sverker Sörlin. 2002. "Narrative and Practice: An Introduction." In *Narrating the Arctic*, ed. M. Bravo and S. Sörlin, 3–33. Canton, MA: Science History Publications.

Brox, Ottar. 1966. *Hva skjer i Nord-Norge? En studie i norsk utkantpolitikk.* (*What Happens in Northern Norway? A Study of Policy for the Periphery*) Oslo: Pax.

Cassidy, Rebecca, and Molly Mullin, eds. 2007. *Where the Wild Things Are Now: Domestication Reconsidered.* Oxford: Berg.

Childe, V. Gordon. 1928. *The Most Ancient East.* London: Kagan Paul.

Clutton-Brock, Juliet. 1994. "The Unnatural World: Behavioral Aspects of Humans and Animals in the Process of Domestication." In *Animals and Human Society*, ed. A. Manning and J. A. Serpell, 23–36. London: Routledge.

Connerton, Paul. 1989. *How Societies Remember.* Cambridge: Cambridge University Press.

Coppinger, Raymond, and Lorna Coppinger. 2001. *Dogs: A Startling New Understanding in Canine Origin, Behaviour and Evolution.* New York: Scribner.

Crosby, Alfred W. 1986. *Ecological Imperialism: The Biological Expansion of Europe, 900–1900.* Cambridge: Cambridge University Press.

de la Cadena, Marisol. 2014. "Runa, But Not Only. " *HAU: Journal of Ethnographic Theory* 4, no. 2: 253–59.

Déscola, Philippe. 2012. "Beyond Nature and Culture: Forms of Attachment." *HAU: Journal of Ethnographic Theory* 2, no. 1: 447–71.

Fageraas, Knut. 2016. "Housing Eiders, Making Heritage: The Changing Context of the Human-Eider Relationship in the Vega Archipelago, Norway." In *Animal Housing*, ed. K. Bjørkdahl and T. Druglitrø, 82–99. London: Routledge.

Fijn, Natasha. 2011. *Living with Herds: Human-Animal Coexistence in Mongolia.* Cambridge: Cambridge University Press.

Fijn, Natasha. 2015. "The Domestic and the Wild in the Mongolian Horse and the Takhi." In *Taxonomic Taperstries: The Threads of Evolutionary, Behavioral and Conservation Research*, ed. A. M. Behie and M. F. Oxenham: Australian National University, Canberra, ANU Press.

Gammage, Bill. 2011. *The Biggest Estate on Earth: How Aborigines Made Australia.* Sydney: Allen and Unwin.

Gifford-Gonzalez, Diane, and Olivier Hanotte. 2011. "Domesticating Animals in Africa: Implications of Genetic and Archaeological Findings." *Journal of World Prehistory* 24, no. 1: 1–23.

Hansen, Lars Ivar, and Bjørnar Olsen. 2004. *Samenes historie fram til 1750*. Oslo: Cappelens Akademisk.

Ingold, Tim. 2000. *The Perception of the Environment*. London: Routledge.

Ingold, Tim, and Gísli Pálsson. 2013. *Biosocial Becomings: Integrating Social and Biological Anthropology*. Cambridge: Cambridge University Press.

Joks, Solveig, and John Law. 2017. "Sámi Salmon, State Salmon: TEK, Technoscience and Care." *The Sociological Review Monographs*, 65, no. 2: 150–71.

Kirksey, Eben, and Stefan Helmreich. 2010. "The Emergence of Multispecies Ethnography." *Cultural Anthropology* 25, no. 3: 545–76.

Kramvig, Britt. 2005. "The Silent Language of Ethnicity." *European Journal for Cultural Studies* 8, no. 1: 45–64.

Law, John, and Marianne E. Lien. 2013. "Slippery: Field Notes on Empirical Ontology." *Social Studies of Science* 43, no. 3: 363–78.

Law, John, and Liv Østmo. 2016. "On Land and Lakes: Colonizing the North," *Technosphere Magazine*. Available at https://technosphere-magazine.hkw.de/p/458a7290-0e3b-11e7-a5d7-f7e271a06d5f.

Lien, Marianne E. 2014. "Fluid Subsistences: Towards a Better Understanding of Northern Livelihoods." *Polar Record* 50, no. 255: 440–41.

Lien, Marianne E. 2015. *Becoming Salmon: Aquaculture and the Domestication of a Fish*. Oakland: University California Press.

Mathews, Andrew. 2017. "Ghostly Forms and Forest Histories." In *Arts of Living on a Damaged Planet*, ed. Anna Tsing, Heather Anne Swanson, Elaine Gan, and Nils Bubandt. Minneapolis: University of Minnesota Press.

Nieminen, Mauri, et al. 2011. "Mortality Rates and Survival among Semi-Domesticated Reindeer (*Rangifer tarandus tarandus L*) Calves in Northern Finland. " *Rangifer* 31, no. 1: 71–84.

Nustad, Knut G. 2015. *Creating Africas; Struggles over Nature, Conservation and Land*. London: Hurst.

O'Connor, Terence P. 1997. "Working at Relationships: Another Look at Animal Domestication." *Antiquity* 71, no. 271: 149–56.

Oma, Kristin Armstrong. 2010. "Between Trust and Domination: Social Contracts between Humans and Animals." *World Archaeology* 42, no. 2: 175–87.

Paine, Robert. 1988. "Reindeer and Caribou: *Rangifer tarandus* in the Wild and under Pastoralism." *Polar Record* 24, no. 148: 31–42.

Rybråten, Stine. 2014. "*This Is not a Wilderness. This Is Where We Live": Enacting Nature in Unjárga/Nesseby, Northern Norway*. PhD thesis, Faculty of Social Science, University of Oslo.

Sara, Mikkel Nils. 2009. "Siida and Traditional Sámi Reindeer Herding Knowledge." *Northern Review* 30: 159–78.

Scott, J. 2011. *The Tanner Lectures on Human Values: Four Domestications; Fire, Plants, Animals, and . . . Us*. http://tannerlectures.utah.edu/_documents/a-to-z/s/Scott_11.pdf: Harvard University.

Shipman, Pat. 2010. "The Animal Connection and Human Evoloution." *Current Anthropology* 51, no. 4: 519–38.

Simon, Scott. 2015. "Real People, Real Dogs, and Pigs for the Ancestors." *American Anthropologist* 117, no. 4: 693–709.
Smith, Bruce D. 2001. "Low-level Food Production." *Journal of Archaeological Research* 9, no. 1: 1–41.
Stépanoff, Charles. 2012. "Human-Animal 'Joint Commitment' in a Reindeer Herding System." *HAU: Journal of Ethnographic Theory* 2, no. 2: 287–312.
Sundsvold, Bente. 2010. "Stedets herligheter—Amenities of Place: Eider Down Harvesting through Changing Times." *Acta Borealia* 27, no. 1: 91–115.
Sundsvold, Bente. 2015. "'Den Nordlandske Fuglepleie': Herligheter, utvær og celeber verdensarv." PhD Dissertation, University of Tromsø.
Swanson, Heather. 2016. "Anthropocene as Political Geology: Current Debates over How to Tell Time." *Science as Culture* 25, no. 1: 157–63.
Tsing, Anna. 2012. "Unruly Edges: Mushrooms as Companion Species." *Environmental Humanities* 1: 141, 54.
Vigne, Jean-Denis. 2011. "The Origins of Animal Domestication and Husbandry: A Major Change in the History of Humanity and the Biosphere." *Comptes Rendus Biologies* 334: 171–81.
Vorren, Ørnulf. 1951. *Reindrift og nomadisme i varangertraktene.* (*Reindeer Herding and Nomadism in Varanger*). Tromsø: Tromsø Museums Årshefter.
Ween, Gro B. 2012. "Resisting the Imminent Death of Wild Salmon: Local Knowledge of Tana Fishermen in Arctic Norway." In *Fishing People of the North: Cultures, Economies, and Management Responding to Change*, ed. C. Carothers, K. Criddle, C. P. Chambers, P. J. Cullenberg, J. Fall, A. Himes-Cornell, J. P. Johnsen, N. Kimball, C. Menzies, and E. S. Springer. Fairbanks: University of Alaska Fairbanks.
Ween, Gro B. 2018. Protection of Sámi intangible cultural heritage and intellectual property rights, and its relation to identity politics in a postcolonial Norway. In C. Antons and W. Logan (eds.) *Intellectual Property, Cultural Property and Intangible Cultural Heritage.* London: Routledge: 157–173.
Ween, Gro B., and Marianne Elisabeth Lien. 2012. "Decolonialization in the Arctic? Nature Practices and Land Rights in the Norwegian High North." *Journal of Rural and Community Development* 7, no. 1: 93–109.
Zeder, Melinda. 2012. "Pathways to Animal Domestication. " In *Biodiversity in Agriculture: Domestication, Evolution, and Sustainability*, ed. P. Gepts, T. R. Famula, R. L. Bettinger, S. B. Brush, and A. B. Damania, 227–59. Cambridge: Cambridge University Press.
Zeder, M., D. Bradley, E. Emshwiller, and B. D. Smith, eds. 2006. *Documenting Domestication: New Genetic and Archeological Paradigms.* Berkeley: University of California Press.

PART II. BEYOND THE FARM

DOMESTICATION AS WORLD-MAKING

6

DOMESTICATION GONE WILD

Pacific Salmon and the Disruption of the Domus · *Heather Anne Swanson*

Where Is the Domus?

All the texts in this volume push us to rethink domestication in one way or another. This chapter's contribution is to disrupt the ways we tend to think about the *geographies* of domestication. If the domus is domestication's space, where *is* the domus? I ask this question through attention to a particular type of creature: Pacific salmon. The transformation of these fish that has occurred through their ocean "ranching" shows how some forms of domestication are not contained on the farm but instead go wild, with rippling effects across wide swaths of landscape. I propose that the question at the heart of this chapter—Where is the salmon domus?—pushes us to rethink our conceptualizations of the process of domestication itself.

My queries about the geographies of salmon domestication have emerged from my ongoing conversations about these fish with Marianne Elisabeth Lien, one of the other editors of this volume. Lien studies domestication from the perspective of Norwegian salmon farms, where Atlantic salmon have been cultivated in net pens since the 1970s.[1] As Lien's work shows, in Norway, the domestication of salmon was originally an attempt to bring fish onto the family farmstead—the classic domus. Norwegian land-based family farmers, seeking to supplement their incomes, were the ones who first began experimenting with salmon cultivation. In this context, it initially seemed self-evident to consider the pens themselves—the rough equivalent of the barn—as the salmon domus, that is, the space of salmon domestication. More recently, however, Lien has asked us to consider expanding our sense of the domus. As salmon farming has expanded and industrialized, it has come to have effects that reach far beyond the borders of the farm. "The salmon domus is more than what meets the eye," Lien writes (2013b). "As a global commodity, feeding on Peruvian anchovietta and destined for Chinese consumers, farmed salmon defies any attempt to pin it down to a particular place."[2]

As Lien points out, Norwegian farmed salmon are reconfiguring other worlds through the connections of political economy. They are remaking coastal ecologies by voraciously consuming feed pellets made from fish caught off South America, and they are remaking middle-class Asian palates as they increasingly appear in stores and restaurants there. The domus stretches outward in other ways, too. Lien illustrates that salmon domestication takes place not only on the farms themselves but also in places like government offices, where fish welfare policies are made.

But how far, exactly, does the salmon domus reach?

The term "domus" is best known as a type of upper-class Roman house whose design enclosed women and slaves within its walls. In *The Domestication of Europe*, Ian Hodder (1990) explores how the domus was not only a physical structure but also a conceptual one—and one with even deeper roots.[3] Beginning in the Neolithic, Hodder argues, the domus comes to stand in opposition to the agrios—the wild or savage—creating the conceptual binaries between culture and nature, as well as domestic and wild, that have become organizing frameworks for much European thought. For early Europeans, domestication became the act of bringing plants and animals across the threshold, from the wild into the space of the human home. Agriculture, for example, was the taming of the wild by converting a patch of it into a cultured, domestic space.[4]

Over the last several decades, however, the nature-culture binary—the division between the domus and the agrios—has been slowly crumbling within the Euro-descendent worlds where it has long held conceptual sway. The recent rise of the term "Anthropocene" is just one example. Coined by Dutch chemist Paul Crutzen in 2000, the Anthropocene is a proposed name for our current geologic epoch, in which humans have become the earth's most significant geologic force.[5] In the Anthropocene, humans are no longer creatures who merely alter regional ecologies; rather, through fossil fuel consumption, global climate change, and mass species extinctions, humans trump even the glaciers in their planetary effects. *Everything*, the Anthropocene concept asserts, is now anthropogenic to some degree. In this way, the Anthropocene disrupts our notion of the domus and becomes a lure for rethinking domestication. If we take seriously the idea of globally pervasive anthropogenic change, then the domus is everywhere.

At the same time, however, the Anthropocene is also a call to recognize that while human influence is ubiquitous, people are thoroughly dependent on multispecies arrangements that far exceed their control.[6] In the midst of anthropogenic change, the world continuously surprises us—for better *and* for worse. Human projects have disastrous unforeseen consequences that wreck ecologies. Yet species also form new alliances in industrial ruins. If we take seriously the notion that humans are never in "control"—that we never make the world as we please—then the agrios, the wild, is as omnipresent in the Anthropocene as the domus. How do we engage "domestication" in a time when we can no longer clearly recognize the boundaries of the domus?

My goal in this chapter is to raise questions of how we might want to use the concept of "domestication" in such worlds of ricocheting anthropogenic effects. Building on Lien's work, which points us toward the importance of tracing networks of political economy, government policy-making, and scientific research beyond the farm, I want to push us to open the borders of domestication even further by adding to the mix close attention to *ecological relations*. Focusing specifically on salmon, I seek to ask: how do the borders of the salmon domus get configured not only by such things as the supply chains of food pellet production but also by the movements of fish themselves?

Pacific Salmon and Hatchery-Based Domestication

To address this question, I begin by shifting the starting point for conversations about salmon domestication from the pen to the North Pacific Ocean. Although North Atlantic salmon farming is only a few decades old, the industrial cultivation of Pacific salmon began much earlier in both the United States and Japan. In 1864, two fishermen from the U.S. state of Maine, the Hume brothers, set up the world's first commercial salmon cannery on the Columbia River, along the border between Oregon and Washington states. Almost immediately, they spawned a new industry. With salmon safely preserved in metal vessels, Columbia River fish could be shipped to markets anywhere in the world. By 1873, customs records show that Columbia River salmon were already being directly exported to Europe and Australia (Penner 2005, 10). In England, canned salmon—a new industrial product—became a key foodstuff for another product of industrialization: the British urban working class. As such markets for canned salmon rapidly expanded, so did the geographical scope of the industry itself. In short order, salmon canneries spread north to Washington state, British Columbia, and Alaska. Just as quickly, they hopped across the ocean to Japan, where the first cannery opened on the island of Hokkaido in 1876.[7]

Salmon *hatcheries* soon followed salmon canneries, as fishermen and business owners sought to augment their supplies of fish. In 1877, the first hatchery appeared in Oregon, followed only one year later by the first facility on Hokkaido. While salmon hatcheries do not keep fish enclosed in pens, they have nonetheless substantially "domesticated" salmon by altering their reproductive processes. Although Pacific salmon form a different genus from Atlantic fish, they share a similar life cycle pattern: they are born in freshwater streams, migrate to the ocean where they feed for one to four years, and then return to the stream of their birth, where they reproduce, typically dying soon thereafter. When they spawn on their own, the survival rate for Pacific salmon from egg to spawning adult is unbelievably low, usually less than 0.1 percent.[8] The greatest numbers of deaths occur among eggs and newly hatched salmon, which are eaten by both birds and larger freshwater fish. Since their late-nineteenth-century invention, salmon hatcheries have aimed to increase the juvenile survival rate and ultimately the overall number of salmon by protecting them from predators (as well as other dangers such as floods) during their early life stages. Hatcheries remove eggs and sperm from mature salmon, then mix them by hand in buckets. The eggs are transferred to incubator trays or bins for their initial development and then into tanks after they hatch. The young salmon are fed pelleted food, until they are large enough to migrate to the ocean. At that time, they are released into a nearby river, which they descend en route to the sea. After maturing, they return to the stream of their birth, where they are either captured by fishermen (becoming a human meal) or by hatchery workers (becoming brood stock for the next generation of hatchery fish). Tellingly, the process is often called "sea ranching" or "ocean ranching"—alluding to its similarities to open-range cattle production.

The direct effects of hatcheries on the salmon reared within its facilities have been profound. As is the case for cattle on the open range, the mobility of hatchery-reared salmon cannot be equated with a relative lack of human influence on their lives. Through alterations of both genetics and early rearing environments, hatcheries have remade Pacific salmon bodies and their rhythms. Some hatcheries have intentionally engaged in selective breeding practices, choosing to spawn the biggest fish that return in the hopes of increasing their genes and thus average fish size. Others have acquired fertilized salmon eggs from areas seen as having more desirable fish, releasing the resulting offspring and their genes into their own rivers to enhance their stocks.[9] In other cases, the modifications of salmon have been more happenstance. For example, hatchery workers often worry that they might not fill their quotas of eggs if they wait until late in the season, so, until the past few years, they consistently used the earliest returning fish as brood stock. As a result, the genes of early returning fish have nearly always been overrepresented, and over the course of many decades, the timing of hatchery salmon runs has crept earlier (Quinn et al. 2002). Fish runs that once peaked in October now hit their zenith in September. Behaviorally, salmon also have been changed. Hatchery salmon are less savvy about predator avoidance than their stream-born peers. For example, some former hatchery salmon tend to confuse the movements of low-flying fish-eating birds with those of the hatchery workers who previously brought them food.

While salmon hatcheries exist in all North Pacific nations, the hatchery production of Pacific salmon has been geographically patchy. Salmon range from northern California to northern Japan and South Korea in a more or less continuous arc, yet hatcheries initially clustered primarily at this arc's ends. In the past several decades, a substantial number of hatcheries have been constructed in parts of Alaska and Russia, but they have been unevenly distributed within those regions, leaving vast tracts hatchery-free.[10] Yet, as I explore in the next section, even those salmon who dwell in these patches, which have so far fallen outside of hatcheries' direct cultivation regimes, have not been unchanged by salmon domestication.

Norton Sound Chum Salmon

The chum salmon (*Oncorhynchus keta*) of Norton Sound, Alaska, are arguably among the most "wild" salmon on earth. They have never been subject to hatchery cultivation or put into pens. Yet they, and the ecologies of their region, are being slowly transformed in ways that ask us to rethink the concepts of domestication itself.

Norton Sound is not an untouched region, nor one where domestication is completely new. Norton Sound's indigenous communities have long made use of its salmon, including as a food source for their domestic dogs. But except for a brief blip in the late 1890s, when explorers found gold on the Seward Peninsula, along the north side of Norton Sound, the area remained on the margins of industrial activity. Commercial fishing became central to Alaska's overall economy but not in Norton Sound. The geology and geography of Norton Sound kept it off trade routes; it was too shallow for early sailing ships to enter, and it was far enough north that the longer journey reduced profits (Norton Sound Steering Committee 2003). Fur traders passed over it, and so did later salmon processors. Alaska had plenty of salmon in areas with lower transportation costs.

While salmon hatcheries proliferated in Japan and other parts of the United States, they remained slow to catch on in Alaska. Alaska was so big and so wild that its salmon seemed inexhaustible—and thus not in need of supplementation. Eventually, however, even Alaska's salmon abundance began to dwindle. In the 1970s, commercial salmon catches plummeted to one-fifth of what they had been a few decades earlier, and suddenly, hatchery production became much more alluring. The facilities soon started cropping up in parts of the state. None of the new hatcheries, however, got sited in Norton Sound. Perhaps they would have if Alaska fisheries officials had opted for a state-funded hatchery model like that employed in Oregon and Washington. Had they done so, there likely would have been political pressure to distribute government funds relatively equitably across the state. But Alaskan officials, inspired by Japanese models of private cooperative hatcheries, encouraged the establishment of similar institutions in their state. The private co-ops made the "smart" business decision to locate their hatcheries in southeast Alaska and Prince William Sound, where the infrastructure for large commercial fisheries was already in place. Norton Sound got passed over, because it had poorer transportation systems, less economic integration, and fewer white commercial fishermen.

Yet, while the Norton Sound chum salmon remained economically marginal, they were pulled into the center of new conversations about biodiversity and genetic diversity. These were, of course, widely circulating discourses in conservation circles in the 1990s. In the context of salmon, they raised new concerns about hatcheries and their rippling effects. Due to the processes of cultivation, hatchery fish did not have either the genetic specificity or the diversity of their stream-spawning relatives. Populations of stream-spawning salmon are genetically and behaviorally distinct because they are adapted to the unique conditions of their rivers. By mixing the genes of fish from different places and rearing them in tanks, hatcheries broke such links between salmon and their specific rivers. Within only a couple of generations of mate selection and captive juvenile rearing, hatchery fish developed more homogenous genetic profiles than stream-spawning salmon, and individual hatchery fish carried genetic signatures that clearly linked them to their facilities.

When scientists began their efforts to catalog the diversity of stream-spawning Pacific salmon, they soon became alarmed. Many of the "wild" populations near hatcheries showed genetic incursion from nearby hatchery stocks. Although salmon generally return to their natal stream—or in the case of hatchery fish, to the entrance of their facility—some fish stray into other waterways. Hatchery fish seemed to stray at a greater rate than other fish, taking their less diverse and less specific gene mixture with them. Because the sheer number of hatchery fish was so high, their strays would sometimes numerically swamp the stream-spawning fish, their industrially shaped genes displacing those adapted to the uniqueness of their places. Scientists themselves rapidly became unsure about the edges of the domus: although they had previously relied on a categorical split between "wild" and "hatchery" fish, reality seemed too slippery for such a binary. Were the majority of nonhatchery salmon "wild" anymore, when they had interbred with hatchery strays and contained genes that were directly traceable to hatchery stocks? Fisheries scientists switched from the term "wild" to that of "stream-spawning" to refer to most nonhatchery salmon to indicate that while such fish were not explicitly cultivated, they were not entirely outside the domus, either.

Precisely because Norton Sound salmon had been left out of hatchery cultivation, they suddenly became interesting. The region's salmon seemed to be among the few truly "wild" fish remaining, and thus some of the most valuable for conservation. Yet even these fish at the edge of the world, scientists soon realized, have gotten caught up in the rippling effects of domestication projects.

Trans-Pacific Domestication?

Today, the North Pacific is flooded with chum salmon—not from Norton Sound but from Hokkaido, Japan. In the mid-nineteenth century, Japanese officials embarked on an ambitious project. Faced with threats of Euro-American colonization, Japanese elites sought to transform their association of feudal dynasties into a modern nation-state. One of their state-making projects was the colonization of Hokkaido. When the Japanese administrators assigned to the task of colonizing Hokkaido heard about the U.S. canned salmon industry, they thought they could establish something similar in northern Japan. If Japan was going to develop national power quickly and forcefully enough to avoid Euro-American domination, they were going to need a favorable balance of trade to acquire foreign currency. They needed new economic formations, and canned salmon seemed promising.

But so did the development of large-scale agriculture in Hokkaido, a process that included draining marshes, channelizing rivers, and constructing dams. Such changes in Hokkaido's landscapes, coupled with the rapid increase in salmon harvests, quickly decimated Hokkaido's salmon numbers. Regional officials, however, were quick to turn to fish hatcheries, which they constructed around the island. Initially, the hatcheries did little to boost salmon populations. Although hatcheries released large numbers of juvenile fish, most of the hatchery smolts died soon after they reached the ocean, contributing almost nothing to overall salmon numbers.

Hatchery failures, however, didn't create much of a stir in Japan. As salmon disappeared from Hokkaido, the island's Japanese fishermen just moved north to the waters around Kamchatka and the Kurils (then under Japanese control), where they found plenty of additional fish. At the end of World War II, however, things abruptly changed. Japan lost control over the islands in the Okhotsk Sea and thus its access to Russian-bound salmon. By the late 1960s, it also became clear that Japanese access to high-seas salmon fishing in the middle of the North Pacific was soon to be eliminated by a series of international treaties.

At this point, restoring Hokkaido's wild salmon populations wasn't a viable option: the island's rivers were far too degraded to support sizable natural stocks. So, the Japanese government decided that they needed better hatcheries to create salmon that they could capture in the waters around Hokkaido. The sea-ranched salmon would do what the Japanese boats could no longer legally do; they would access the resources of the high seas, where they would feed, before returning to the rivers of their birth, where they would be harvested. The Japanese government poured money into hatchery research and development, and by the mid-1970s, this investment started to generate a visible return. Between 1970 and 2000, the number of these ranched fish increased tenfold. Overall salmon numbers skyrocketed to more than five times their estimated pre-1800 abundance; and more than 95 percent of these fish came from hatcheries. Today, Japanese hatchery chum salmon, approximately two billion of whom are released every year (Ruggerone et al. 2010), are "the dominant chum salmon stock in the Bering Sea and North Pacific Ocean" (Urawa et al. 2009).

The influence of those fish, however, is not even across this oceanic region. Chum salmon have long migratory routes that they follow to pockets of the North Pacific and Bering Sea. Different salmon populations make their way to different patches. Yet, despite their origins on different continents, Japanese hatchery salmon and Norton Sound fish appear to hang out in almost exactly the same neighborhoods, and their overlapping ranges seem to have consequences for all involved. With the influx of a large number of hatchery salmon, the competition for prey in these areas has grown much greater. As the fish increasingly struggle to find adequate nutrition, average body size for fish from both Japanese hatcheries and Norton Sound is shrinking.[11] This matters, because when salmon have smaller body sizes, female fish produce fewer eggs when they spawn. The decrease in eggs isn't much of a problem for hatcheries, where reproduction is managed to insure that nearly 100 percent of eggs are fertilized and where, protected from the fluxes of life in a river, about 98 percent of fish survive into young adulthood. For the Norton Sound salmon, however, such changes are potentially catastrophic. In a stream, it's not uncommon for only a few percent of salmon eggs to become out-migrating juvenile fish and for only about 1.8 percent of those out-migrating juvenile fish to eventually return to spawn (Kaeriyama et al. 2007). When survival rates are that small, even a slight decline in the number of eggs per female fish can drop the numbers of returning adults low enough to turn a thriving salmon population into a dwindling one.

The best available evidence strongly suggests that this is what is happening to the Norton Sound salmon. The abundance of Norton Sound fish has declined since the 1980s, and several populations have been classified as "Stocks of Concern" by the state of Alaska (Ruggerone et al. 2011). As Japanese hatchery salmon numbers have gone up, Norton Sound populations have gone down—replacing the North Pacific's "wild" salmon with hatchery ones (Ruggerone et al. 2011).

The effects of such a change are not entirely known. But it is not unreasonable to speculate that it has far-reaching implications for landscapes and oceanscapes around the North Pacific. As they peregrinate, Japanese hatchery chum salmon consume large quantities of squid larvae, crustaceans, and small fish. What effects might their intense feeding have not only on those species themselves but also on the many ocean organisms that share these prey? The effects of Japanese hatchery salmon likely spread into the watersheds of Alaska, as well. When they return to spawn and die, salmon bring significant quantities of marine-derived phosphorous and nitrogen back to streams. Numerous scientific studies have shown that nutrients from salmon carcasses affect vegetation patterns (Helfield and Naiman 2001), invertebrate densities (Hocking and Reimchen 2002), and even songbird populations (Christie and Reimchen 2008). Salmon-derived nutrients can be clearly identified in the wood of trees more than five hundred meters from salmon-bearing waterways (Rozell 2004). Thus, when salmon populations (like those in Norton Sound) decline, it alters the region's fundamental nutrient cycling patterns in unpredictable ways.

Domestic Disturbances: Rethinking the Edges of the Domus

Where, then, do we draw borders around the domus? In light of the story above, I am clearly not asking a question of whether or not we classify an *individual* organism as domestic or wild. Rather, I am asking us to reconsider how we tell stories about domestication's wide-ranging effects. Do we consistently add Peruvian anchoveta into our stories? What about changes in North Pacific krill populations due to Japanese hatchery salmon? What about the anthropogenic climate change that may also be affecting Norton Sound fish? Where and how do we want to make such cuts? Just as importantly, if one *wants* Peru or the North Pacific inside stories about domestication, how does one get them there?

My guide here is Donna Haraway, whose work on dogs begins within her own middle-class American domus but does not stay within its walls. When Haraway engages in the simple act of touching her dog, the histories of domestication that they have both inherited open outward into multispecies landscapes (Haraway 2003, 2008). For Haraway, the scenes of dog domestication are not limited to one-on-one interactions between dogs and people; rather, they include multispecies formations, such as the emergence of her own herding dog, an Australian shepherd, within the settler colonial projects that remade Australian ecologies. How does the domus—and thus domestication—look different if, following Haraway, we allow it to exceed the human household with its kennel? This is the kind of approach that I suggest we take with salmon—a shift in our scene of analysis from the pen to *landscapes and oceanscapes.*

Typically, species dyads have been at the center of conversations about domestication. Species such as horses, cattle, or salmon get domesticated, one at a time, in relation to humans. Of course, we are all aware that practices of domestication produce rippling ecological effects: cattle rearing changes grasslands, while salmon farming requires the harvest of small fish to produce pelleted feed. However, such processes have generally been seen as external to acts of domestication themselves. But what if we place landscape changes at the heart of domestication stories, exploring how domestication itself is best understood not as a new relation between humans and a single kind of plant or animal but as a reconfiguration of multispecies assemblages?

Quite a few scholars have talked about the "domestication" of landscapes. But such uses of domestication have often been dismissed as too metaphorical. In the introduction to *Where the Wild Things Are Now* (2007), a previous attempt to reconsider domestication in anthropology, the volume's editors explicitly describe themselves as "resistant" to broad usage of the term, including its application to landscape: "What does it mean to domesticate desire? Or Europe? Or death? Or mobile telephones? What does it mean to describe people as undergoing a process of domestication? . . . In these cases, domestication seems an attractive but ultimately unsatisfying euphemism, with a catch-all property that makes critique impossible" (Cassidy and Mullins 2007, 3). The "domestication" of the North Pacific oceanscapes and Alaskan landscapes, however, is not mere metaphor. Rather, it is a reverberation of domestication in its most narrow sense. The changes that we see in these landscapes are materially tied to the production of hatchery salmon. Instead of making critique impossible (as Cassidy and Mullins claim), I argue that it opens up new, and potentially even more powerful, forms of critique.

For an example beyond salmon, consider the work of my colleague Katy Overstreet, whose research focuses on dairy cattle in the U.S. state of Wisconsin (unpublished dissertation manuscript). One approach to domestication here would be to look at how cow bodies have been and continue to be reshaped by the demands of milk production. Another would be to look at how human lifeways, such as those of the dairy farmers, have come into being through their relations with cattle. Overstreet certainly examines such things. However, she also does something more: She traces how dairy production reverberates through webs of multispecies relations. Concretely, she has examined how the mono-crop fields used to produce animal fodder have altered insect communities and how changes in the layouts of dairy farmsteads have affected bird life. She connects these outer landscapes to inner ones, asking how the gut bacteria that reside in ruminant stomachs have also been changed by shifts in the worlds that cows and people participate in making. Overstreet's work positions such changes not as mere "effects" of domestication, but as part and parcel of domestication itself. The object of her domestication story is not cows, per se, but Wisconsin landscapes. Such a shift in objects does not shut down critique around topics such as cow welfare or the ethics of forced industrial lactation, which remain central to Overstreet's work. Rather, they add a new set of questions about "caring for country" into the mix (Rose 2004). Thinking through landscapes forces us to bring new beings—for example, bees who are killed by pesticides applied to cattle feed crops—into our analyses of domestication.

Domestication as Disorientation

When one takes landscapes or oceanscapes as a unit of analysis, it changes our sense of what domestication does. Landscapes show us that even when domestic creatures are kept more-or-less contained, the effects of their domestication is not. Outside the pen, we see that domestication goes *wild*, altering worlds in all kinds of unexpected ways. I draw here on anthropologist Deborah Bird Rose's use of the term "wildness" (Rose 2004). For Rose and her Aboriginal Australian collaborators, "the wild" does not refer to a pristine, untouched space. Rather, wildness is something made through the violences of colonial settlement. "Wild people (colonizers) make wild country (degraded and failing)," Rose writes (2004, 4). And, in nineteenth- and twentieth-century Australia, they did so through their relations with domestic animals. As cattle trampled and overgrazed the Australian outback, they turned it into a crazy place of ruins. "This 'wild,'" Rose explains, "was a place where the life of the country was falling down into the gullies and washing away with the rains" (2004:4). In this sense, hatchery salmon have made Norton Sound more—not less—wild.

This wildness of domestication, I argue, emerges from the way domestication creates disorientation in relation to landscapes. Indeed, I propose that we might reimagine *domestication as a process of disorientation*, a process of disrupting humans' and nonhumans' ties to landscapes to make the domus central to their worlds. Such disorientation takes many forms. For just one example, consider domestic dogs, who now turn toward human handlers for food and companionship, in counterpoint to their wolf ancestors, who turned toward packs, paths of movement, and prey. Such disorientations may bring some pleasures within the home, that is, the joy of play that one sees between owners and their dogs; but they are also undoubtedly disruptive both within and beyond the domus.[12]

Let us return for a moment to the Australian outback from which Rose writes. It is a place whose ruination is the product of both Euro-Americans who are unattuned to landscapes and the animals whom they have made similarly so. People deaf to the multispecies relations of such a place and bent on conquest brought cattle, bred for meat production, who were similarly poorly oriented to the continent's landscapes. They stripped the ground of its vegetation, while the pounding of their hooves damaged the soil. Domestication, here, disoriented not only particular beings but also *entire landscapes.* Once the cattle and colonists came, nothing—not native plants, not endemic animals, not Aboriginal peoples—remained unchanged in domestication's wild cascades. Species declines and destruction of lifeways proliferated as large swaths of Australia were disoriented into "rangelands."

Even when domestication leads to seemingly "sustainable" worlds, it still involves disorientation. Consider the case of hefted sheep. In the unfenced highlands of Britain's Lake District, farmers work to "bind" domesticated sheep flocks to particular patches of land (see Olwig 2013). Initially, the sheep must be prevented from wandering away by constant shepherding. Yet over time, the animals begin to bond, or heft, to their land, developing what one might call a sense of place. Once a flock is hefted, they remain so, passing their knowledge of how to navigate the area's grazing and sheltering spaces from ewe to lamb across many generations. In one sense, hefting seems like a reorientation, not a disorientation; it seems like a case where domestic animals successfully move further out from the domus proper by integrating into heathland ecologies. Yet ostensible reorientation of hefted sheep only becomes possible in the midst of disorientations. For example, the binding of sheep to the land has required the elimination of wolves from the region and the utter disruption of ecologies attuned to predators. The result is a disoriented landscape, a wild landscape.

Humans have not escaped the disorientation of the home-domus-domestication trifecta. Over time, we have become disoriented toward cars, single family dwellings, cheap factory-farmed foodstuffs, McDonalds, and watching TV on the couch. Our bodies are increasingly disoriented by refined grains and hormone-laden meat products that change our gut bacteria, leading us to crave more of the same in even larger quantities. Most importantly, however, domestication has disoriented us by leading us to think that we are OK without landscapes—to think that *home* is enough. The domus and its hubris of human control has led us to imagine ourselves as separate from the multispecies relations of the agrios, allowing us to forget our relentless entanglements with the landscapes. Inside the domus, we become inured to forgetting about the world. As we become unable to notice our relations to landscapes, we become disoriented to them ourselves.

All these forms of disorientation come into play when we think about salmon. Salmon are typically called "homing" fish because they more or less faithfully return to spawn in their natal streams, which have been widely construed by Euro-Americans, including biologists, as salmon homes.[13] Fish hatcheries, the technology of Pacific salmon domestication, have relied on the logic that one can substitute one home (the hatchery/human domus) for another (the river). Salmon, however, do not have a "home" as such; rather, they navigate a series of waterscapes, including diverse parts of rivers, estuaries, and oceans. Indeed, attempts at home-ing (as in domestication, or the bringing of salmon into the domus) has literally messed with salmon homing of another kind—namely, the ability of these fish to negotiate waterscapes and their multispecies relations. When Pacific salmon swim in the open ocean, they navigate in complex ways that are not completely understood but that clearly rely, at least in part, on sensing the earth's magnetic field (Putman et al. 2013). When salmon were experimentally exposed to very weak magnetic fields—far less than those required to move a compass needle—they oriented themselves in relation to the fields within minutes (Oregon State University 2014a). The problem, however, is that hatchery rearing seems to disrupt the ability of salmon to develop their internal compass. Because the earth's magnetic field is so weak, salmon must be incredibly sensitive to small magnetic differences—meaning that it doesn't take much to mess with their map sense.

Hatcheries, however, are filled with metal—pipes, electrical wires, reinforcing iron, and steel rebar—that create magnetic interference and throw off the orientational development of young fish (Putman et al. 2014; Oregon State University 2014b). The disorientation of salmon that emerges from their domestication makes once familiar oceans wild places for them. When they can't navigate well, salmon cannot locate the best feeding grounds, making survival much more of a challenge. Put in another way, *home makes the wild* not only through binary opposition but also through material practices. Replacing hatcheries' metal pipes and equipment with nonmagnetic materials, such as fiberglass and plastic, have been offered up as possible solutions (Oregon State University 2014b). But a simple change of materials seems to be no panacea. When both stream-born and hatchery juvenile fish migrate downstream, their journeys often traverse metallized waterscapes. It is likely, although as yet unproven, that the hydroelectric dams through which many salmon must pass en route to the ocean also contribute to the magnetic disorientation of migrating fish (Oregon State University 2014a).[14]

Theory and Politics beyond the Pen: A Tentative Conclusion

Our ideas about the domus have limited where we look for domestication. We expect to find domestication in contained spaces—in both our literal and metaphorical backyards. But when we focus on the scenes that seem "domestic," we miss how domestication goes wild, reverberating across landscapes. In this final section, I ask how a shift from seeing domestication as a project of enclosure to an act of landscape disorientation raises different theoretical and political possibilities that might further enrich conversations around domestication.

As feminists and animal rights activists have long shown through their challenges to the confinement of women and animals, domestication *is* political. Yet such politics have too often confined themselves to the domestic—the scene of the home or the pen. While it addresses different concerns, most domestication work in cultural anthropology and science and technology studies has developed its politics from similar narrow scenes. By presenting domestication as an always reciprocal, even if uneven, process, such scholarship has undoubtedly made important contributions to rethinking intentionality, agency, and self/other binaries—all politically potent interventions. But to date, both the theory and the politics of this domestication scholarship has stayed mostly on the farm, without venturing too much into landscapes.

I have tried to allow this chapter to emerge from a different set of politics: those of the Anthropocene. Through its focus on widespread ecological crisis, the Anthropocene draws us directly into multispecies landscapes by pushing us to think about the multiplying human effects on them. But in their efforts to stress the ubiquity of anthropogenic change, Anthropocene scholars have thought little about how best to use the concept of domestication within their politics for living on a damaged planet. This chapter has played with one possible way to do so, by making landscapes the primary unit through which we view the practices of domestication and their effects.

Focusing on landscapes does many things, both theoretically and politically. First, it brings political economy *inside* domestication itself. If we ask about domestication with an eye toward landscapes, we cannot *not* ask about the Peruvian anchovietta. Just as importantly, focusing on landscapes prompts us to move beyond supply chains to attend to effects that move through nonhumans. Such a change has all kinds of political implications, as it demands that we engage with scientists in new ways to trace such effects. Lastly, landscapes hail us to see domestication as disorientation—of both individuals and multispecies communities. In this way, we come to see domestication as a disruption of landscape "response-ability," in Karen Barad's sense of the term (2007)—both animals and people lose their ability to respond well to the worlds around them. When they become insensitive, they cannot engage in the common politics—those that are not based on a nature-culture divide—that are needed to face the messes of the earth (Latour 2013).

How, this chapter has asked, might we better tell stories of salmon domestication—stories that reach out to Norton Sound? I have proposed that the answer lies in thinking from and with landscapes.

NOTES

1. See Lien and Law (2011).

2. Lien extends this idea in another paper (2013a): "A considerable part of the feed pellets which are fed to salmon in Norway is sourced from fisheries in the South Pacific, and from anchovietta, which has been key source of subsistence for Peruvian fishermen for generations. Now they supply fish to fish pellet processing plants. Are these fishermen domesticated too? And what about the seascapes of the South Pacific?"

3. Hodder, however, also traces the origin of the word to other languages, including Sanskrit (*damas*), Old Slavonic (*domu*), Old Irish (*doim*), and the Indo-European (*dom- or dem-*) (Hodder 1990, 45).

4. Hodder defines the domus as "the concept and practice of nurturing and caring, but at a still more general level it obtains its dramatic force from the exclusion, control, and domination of the wild, the outside (which I shall later describe as the 'agrios'). Culture, then, is opposed to nature, but in a historically specific manner" (1990, 45).

5. See Steffen et al. 2011.

6. See Tsing (2015).

7. See Swanson (2013).

8. See http://www.wdfw.wa.gov/fishing/salmon/chum/life_history/index.html.

9. See Taylor (1999).

10. Funding, local political clout, the amenability of different salmon species to cultivation, the ability to get fish to market, and desires to protect wild fish diversity have all influenced the siting of hatcheries.

11. In addition, as hatchery production has ramped up, the age of sexual maturity for both hatchery and stream-spawning salmon has risen.

12. Melinda Zeder (2012) also describes how domestication disorients animals by "selecting for reduced wariness and low reactivity" (232), as well as by causing reductions in brain size and function. These changes are not trivial, with brain size reductions typically ranging from about 14 to 33 percent (2012, 233).

13. For example, Quinn n.d. states, "For a wild fish, home is the natal stream where it incubated, hatched, and emerged. Home is thus, essentially, the redd."

14. It is worth noting here that hatcheries and dams are deeply intertwined, especially in Oregon and Washington, where hatchery production of salmon helped to legitimate dam construction and where, conversely, dam construction served as an impetus for more hatchery construction (Taylor 1999). Dams, too, thus seem to belong inside some salmon domestication stories.

REFERENCES

Barad, Karen. 2007. *Meeting the Universe Halfway: Quantum Physics and the Entanglement of Matter and Meaning.* Durham: Duke University Press.

Cassidy, Rebecca, and Molly Mullin, eds. 2007. *Where the Wild Things Are Now: Domestication Reconsidered.* Oxford: Berg.

Christie, Katie S., and Thomas E. Reimchen. 2008. "Presence of Salmon Increases Passerine Density on Pacific Northwest Streams." *Auk* 125: 51–59.

Haraway, Donna J. 2003. *The Companion Species Manifesto: Dogs, People, and Significant Otherness.* Chicago: Prickly Paradigm.

Haraway, Donna J. 2008. *When Species Meet.* Minneapolis: University of Minnesota Press.

Helfield, James M., and Robert J. Naiman. 2001. "Effects of Salmon-derived Nitrogen on Riparian Forest Growth and Implications for Stream Productivity." *Ecology* 82, no. 9: 2403–09.

Hocking, Morgan D., and Thomas E. Reimchen. 2002. "Salmon-derived Nitrogen in Terrestrial Invertebrates from Coniferous Forests of the Pacific Northwest." *BMC Ecology* 2, no. 4.

Hodder, Ian. 1990. *The Domestication of Europe.* Oxford: Basil Blackwell.

Kaeriyama, Masahide, Akihiko Yatsu, Hideaki Kudo, and Seiichi Saitoh. 2007. "Where, When, and How Does Mortality Occur for Juvenile Chum Salmon *Oncorhynchus keta* in Their First Ocean Year?" *North Pacific Anadromous Fish Commission Technical Report* 7: 52–55.

Latour, Bruno. 2013. "Facing Gaia: Six Lectures on the Political Theology of Nature." Gifford Lectures on Natural Religion, Edinburgh.

Lien, Marianne E. 2013a. "Domestication as Partial Relations; Lively Attachments and the Anthropos of Anthropology." Paper presented at the Sawyer Seminar workshop, University of California, Davis, June 13–14, 2013.

Lien, Marianne E. 2013b. "Salmon Multiple: Creating Dialogue across Disciplinary Boundaries." Keynote presented at the Aarhus University Research on the Anthropocene opening conference. October 6, 2013.

Lien, Marianne E., and John Law. 2011. "'Emergent Aliens': On Salmon, Nature, and Their Enactment." *Ethnos* 76, no. 1: 65–87.

Norton Sound Steering Committee. 2003. "Research and Restoration Plan for Norton Sound Salmon." Accessed November 2, 2014. www.aykssi.org/wp-content/uploads/NS-RR-Plan-rev.pdf.

Olwig, Kenneth R. 2013. "Heidegger, Latour and the Reification of Things: The Inversion and Spatial Enclosure of the Substantive Landscape of Things—The Lake District Case." *Geografiska Annaler: Series B, Human Geography* 95, no. 3: 251–73.

Oregon State University. 2014a. "Study Confirms Link between Salmon Migration and Magnetic Field: News and Research Communications." February 6, 2014. Available online at: http://oregonstate.edu/ua/ncs/archives/2014/feb/study-confirms-link-between-salmon-migration-and-magnetic-field.

Oregon State University. 2014b. "Iron, Steel in Hatcheries may Distort Magnetic 'Map Sense' of Steelhead." *News and Research Communications.* June 3, 2014. Available online at: http://oregonstate.edu/ua/ncs/archives/2014/jun/iron-steel-hatcheries-may-distort-magnetic-"map-sense"-steelhead.

Penner, Liisa. 2005. *Salmon Fever, River's End: Tragedies on the Lower Columbia River in the 1870s, 1880s, and 1890s; Articles from Astoria Newspapers.* Portland: Frank Amato Publications.

Putman, Nathan F., Kenneth J. Lohmann, Emily M. Putman, Thomas P. Quinn, A. Peter Klimley, and David L. G. Noakes. 2013. "Evidence for Geomagnetic Imprinting as a Homing Mechanism in Pacific Salmon." *Current Biology* 23, no. 4: 312–16.

Putman, Nathan F., Amanda M. Meinke, and David L. G. Noakes. 2014. "Rearing in a Distorted Magnetic Field Disrupts the 'Map Sense' of Juvenile Steelhead Trout." *Biology Letters* 10, no. 6: 20140169.

Quinn, Thomas P. n.d. "Homing, Straying, and Colonization." NOAA Tech Memo NMFS NWFSC-30: Genetic Effects of Straying of Non-Native Hatchery Fish into Natural Populations. http://www.nwfsc.noaa.gov/publications/scipubs/techmemos/tm30/quinn.html.

Quinn, Thomas P., Jeramie A. Peterson, Vincent F. Gallucci, William K. Hershberger, and Ernest L. Brannon. 2002. "Artificial Selection and Environmental Change: Countervailing Factors Affecting the Timing of Spawning by Coho and Chinook Salmon." *Transactions of the American Fisheries Society* 131, no. 4: 591–98.

Rose, Deborah Bird. 2004. *Reports from a Wild Country.* Sydney: University of New South Wales Press.

Rozell, Ned. 2004. "Salmon Nose Deep into Alaska Ecosystems." *Alaska Science Forum*, article 1721, October 21.

Ruggerone, Gregory T., Beverly A. Agler, and Jennifer L. Nielsen. 2011. "Evidence for Competition at Sea between Norton Sound Chum Salmon and Asian Hatchery Chum Salmon." *Environmental Biology of Fishes* 94, no. 1: 149–63.

Ruggerone, Gregory T., Randall M. Peterman, Brigitte Dorner, and Katherine W. Myers. 2010. "Magnitude and Trends in Abundance of Hatchery and Wild Pink Salmon, Chum Salmon, and Sockeye Salmon in the North Pacific Ocean." *Marine and Coastal Fisheries* 2, no. 1: 306–28.

Steffen, Will, Jacques Grinevald, Paul Crutzen, and John McNeill. 2011. "The Anthropocene: Conceptual and Historical Perspectives." *Philosophical Transactions of the Royal Society A: Mathematical, Physical, and Engineering Sciences* 369 (1938): 842–67.

Swanson, Heather Anne. 2013. "Caught in Comparisons: Japanese Salmon in an Uneven World." PhD dissertation, University of California, Santa Cruz.

Taylor, Joseph E., III. 1999. *Making Salmon: An Environmental History of the Northwest Fisheries Crisis.* Seattle: University of Washington Press.

Tsing, Anna, 2015. *The Mushroom at the End of the World: On the Possibility of Life in Capitalist Ruins.* Princeton: Princeton University Press.

Urawa, S., S. Sato, P. A. Crane, B. Agler, R. Josephson, and T. Azumaya. 2009. "Stock-Specific Ocean Distribution and Migration of Chum Salmon in the Bering Sea and North Pacific Ocean." *North Pacific Anadromous Fish Commission Bulletin* 5: 131–46.

Zeder, Melinda. 2012. "Pathways to Animal Domestication." In *Biodiversity in Agriculture: Domestication, Evolution, and Sustainability*, ed. P. Gepts, T. R. Famula, R. L. Bettinger et al., 227–59. Cambridge: Cambridge University Press.

7

NATURAL GOODS ON THE FRUIT FRONTIER

Cultivating Apples in Norway · *Frida Hastrup*

The flyer in my hand opens with the headline "The unique taste of Hardanger" and a speech balloon reading, "The best apple in the world." It describes the project known as *Dyrk Smart*, a campaign to present the charms and potentials of fruit production launched by the dominant cooperative fruit packing and storage organizations in Ullensvang municipality in the heart of the Norwegian apple country, by the Hardangerfjord in Western Norway. The flyer goes on to explain that Hardanger, where fruit trees have been tended to since medieval times, is located at the northernmost limit of commercial apple production; yet, "in spite of this," the flyer proclaims, "the conditions are perfect for producing fruit."[1]

I am instantly intrigued by the implication that defying the climatic conditions that make fruit production all but impossible is also seemingly what makes the fruit better. Explicit references to the region as, on the one hand, being located on the limit of the feasible in terms of fruit cultivation and, on the other, having produced perfect apples for centuries make for an interesting kind of fruit frontier—both assertive and noncolonial. What strikes me is how adverse and perfect environmental conditions team up in this case to make a particular natural resource, thus providing an interesting entry point into practices of domestication on the margins. The initiative *Dyrk Smart*—Norwegian for "grow/produce smartly"—holds more puzzles: even though perfect production conditions make orchards an obvious business opportunity in the area, a campaign is necessary to make people in the area, particularly the young, want to enter the commercial fruit business. People I meet at the local packing and storage facility in the village of Lofthus in Ullensvang tell me, with concern, that the local fruit production is a traditional trade that ages too fast. They fear that unless they concentrate on recruiting new producers, they will be put out of business. But according to the campaigners, the time, place, and rewards are just right for this kind of local production. "Farmers will save the world," declares another headline on the *Dyrk Smart* flyer. Under this caption, the renowned trend forecaster Lidewij Edelkoort is cited as stating that people are fed up with bad, unrecognizable food, unhealthy medicine, and toxic ingredients in cosmetic products and whatever other modified and synthetic items the twenty-first century has to offer. People now turn to farmers for natural products, I read. "The farmers are the future. They will feed, dress, medicate and heal us. Ninety percent of all the products we need come from farmers. This is not just a trend, it is a whole movement," the flyer quotes Edelkoort, asserting that for forty years the trend forecaster's predictions have largely been accurate. The origin and subject of the cited forecasts are unclear in the flyer, just as the farmers remain unspecified. But for the *Dyrk Smart* campaigners, the statement is sufficiently specific for it to work as evidence for an increasing demand for locally produced farmers' goods from the Hardanger region—to the benefit of producers and the globe alike. The market for fruit from the area will only grow, they say, making it hard to keep up with the demand—and making it attractive to try. Food scandals and dangerous additives lurking in the food we eat, the flyer further explains, make the presumably cleaner Norwegian fruit people's first choice, even if it comes at a greater cost. Money is no longer people's prime concern, which, interestingly, becomes the main argument for attracting new fruit producers: there is money to be made from the Norwegian orchards. Taken together, there are all the reasons in the world to smartly produce and market cultivated apples—fruits of the *Malus domestica* tree—in the Hardanger region. Why on earth, the campaigners seem to ask, would people not want to pick such low-hanging fruit?

In this contribution, I explore practices and stories of apple production at its northern limit as an intriguing kind of frontier work that produces a particular homeland for the Norwegian fruit—the best in the world. I use the notion of frontier work to point to the series of pioneering efforts to implant and consolidate the Norwegian apples in a specific border landscape, which is seen both as naturally and historically conducive to local quality fruit production and as a scene for active interventions to cultivate these apples. I approach the Norwegian orchards as sites for attempts to coordinate relations—between these apples and other apples, between then and now, between Hardanger and other places, between this market and other markets, and so on. I suggest that these practices of coordination on the fruit frontier that produce Norwegian apples as the best in the world perform an interesting kind of domestication work, not so much because they transform a wild counterpart into a humanly mastered object through conquest but because they craft a specific and situated apple domus through various relations to other apple locales and stories. The production of Norwegian apples, then, is relational work all the way through, not only along the line of wild versus tamed but as an active ordering of specific places in the world and their natural products. The fruit frontier that I evoke here—to begin with simply drawn from my field material, which speaks of northernmost limits, interventions into the ecology meant to warrant proper use of the landscape, and opportunities for growing perfect fruit—is not the gangster fantasy of an up-for-grabs empty and wild resource landscape that Anna Tsing describes so well in her work on Indonesian forests (Tsing 2005, 25 ff.). The apples trees here are not tools for colonization. Rather, the Hardanger fruit frontier is a scene for pioneering attempts to coordinate what is in and what is out of the apple domus, thereby trying to ensure a comparative advantage of Norwegian apples vis-à-vis all other apples in the world. Working this fruit frontier is a matter not of annexation but of establishing and upholding a domus. The peripheral character of the Hardanger region in a world geography of fruit makes it a means of relative distinction and a tool for implicit comparison—in spite of and due to demanding yet perfect climatic conditions for growth (cf. Abram and Lien 2011). As such, this particular fruit frontier becomes a resource through which fruit producers, local historians, campaign managers, anthropologists, and other interested parties can explore the crafting and marketing of local resources in a context of intense global integration and ecological stress.

Viewing apple production in Hardanger as practices that coordinate relations to make a particular frontier for cultivating the best apples in the world, I basically ask how the Norwegian apples become *worldly,* to borrow a phrase from Donna Haraway (2008). I probe this question through conversations during fieldwork in the village of Lofthus in Hardanger in the summer of 2014 and in dialogue with different literary sources. My argument is structured as follows: First, I look at the creative and deliberate *forging of connections* in time and space that work and have worked to implant apples in the landscape and cultural history of the Hardanger area, thereby making them naturally at home there. Second, I focus on the equally creative *cutting of associations* that single out the Norwegian apples as a particular world-class natural good, belonging to a particular encircled land. I then discuss how both sets of (essentially political) fruit frontier practices—the forging of connections that roots local apples *in* Hardanger and the cutting of associations that roots *for* them in the race against other apples and markets—work to make a Norwegian apple domus by deciding what is in and what is out. In conclusion, I reflect on the implications for how we can think about domestication and resources in what Haraway refers to as the "capitalocene" era.

Forging Connections: All Quiet on the Western Front?

"Really, we're at the northern margin of apple production here. But the Gulf Stream keeps the Western fjords warm enough and free of ice, and the mountains provide shelter for the orchard slopes facing the water. . . . Soil conditions, with lots of organic material in it, combined with the many hours of sunlight in the summer and relatively low temperatures, especially during the night, actually make for perfect natural conditions. The apples ripen slowly here, they remain crispy and grow a thin skin, and have just the right balance of acidity and sweetness."

This is how Eivind, a researcher from the fruit and berry section of Bioforsk —the Norwegian Institute for Agricultural and Environmental Research, which is located in the village of Lofthus—frames apple production at sixty degrees north. Intrinsic sensory qualities, it seems, are produced by a distinct ecology, terroir style (Paxson 2010). In Lofthus, though, people do not rely only on inherent quality. In the village, about one hundred producers are organized in a membership-based local packing and storage cooperative—one of the organizations behind the *Dyrk Smart* project—which is mandated to assess fruit quality, pack and store fruit, gradually release it to the market, and not least negotiate prices with retailers on behalf of the Lofthus producers. Most of these are small-scale and part-time producers; fruit production is the sole occupation for no more than one in ten of the orchard owners, although these 10 percent produce about half of the village's yield because of their more efficient and professional production techniques and the additional time they invest in cultivation. The people I meet around town speak of the importance of maintaining a diverse fruit-producing community that benefits from the varied perspectives and experiences of the many growers. Johan, one of the full-time producers, relates the diversity of people involved in fruit directly to the survival of Lofthus as a kind of apple stronghold; fewer orchard owners could more easily have been toppled. At the same time, though, Johan says with a smile that he often looks longingly at his neighbors' sunny slopes, back to back with his own. Indeed, these acres would come in handy for him and his efficient production methods. All in all, though, according to Johan, connecting with a whole community of fruit producers organized in a cooperative makes for better and more robust Norwegian apple land—which, I guess, is only right for a place with perfect climatic conditions.

From a small billboard lining the so-called Fruit Trail that guides trekkers past all the orchard highlights of the village, I learn that the unique production conditions for the local fruit "cannot be imitated naturally." I wonder what to make of this enigmatic statement. Can the conditions be imitated artificially? And what would imitation mean? Or is this statement simply a subtle indication that this landscape for growing apples, albeit demanding, is a special and chosen one? What kind of apple home is this? Natural or artificial landscapes notwithstanding, I read on about centuries of concerted efforts to grow fruit in the area. By 2014, as Eivind of Bioforsk also tells me, Norwegian state politics play a big role through subsidies in ensuring that the small-scale producers in Hardanger can continue their production, which today accounts for between 30 and 40 percent of the country's total fruit production (Syberg 2007, 212). But long before that, clergy in the area took it upon themselves to cultivate apples on the hills facing the fjords, gradually bringing the perfect production conditions into their own. Cistercian monasteries in the region, from the twelfth century onward, were the first to systematically cultivate apples and other fruits and vegetables in the Hardanger area. During my fieldwork, the people I meet between the neat rows of apple trees or in offices refer to this long history and monastic origin of fruit cultivation. On further research, I learn that the Lyse Abbey, founded in 1146 and eventually one of the wealthiest monasteries in Norway, established a plantation in Lofthus that has since been included in the Opedal orchard, the largest estate in the village today (Olafsen 1898, 6; Åsen 2009, 232). In his treatise on fruit cultivation and gardening in medieval Norway, priest, local historian, and garden enthusiast Olaf Olafsen, who was based in Ullensvang from the end of the nineteenth century, writes: "Our historians assume that fruit cultivation and gardening from other countries, that is England, Denmark, Germany and France among others, have spread to the Norwegian monasteries, and that these pursuits of peace and quiet have found fertile ground within the monastic walls" (Olafsen 1898, 8).[2]

The peaceful cultivation of orchard produce that took root in the fertile grounds around the Norwegian monasteries was highly influenced by connections with the world outside the Nordic friars' compounds. From the outset, producing the best apples in the world on the unique Hardanger slopes has been a transnational endeavor that allowed some cultivation components to enter the fruitful Norwegian properties.

In the course of the late eighteenth century, fruit production in the region became more firmly rooted, not least, again, courtesy of the clergy who lived and worked around Hardanger, some of whom brought apple varieties from Denmark and Sweden, such as Gravenstein and Red Aroma, which are still among the dominant, now local, varieties in the area. A reverend by the name Glahn, who was stationed in Ullensvang from 1771, reportedly distributed free scions from suitable trees in the rectory's garden and is credited with having taught the local farmers to graft. This skill of vegetative formation—a coordination of relations, indeed—caused a gradual proliferation of new and sturdier apple varieties, often procured from abroad, to grow in Hardanger (Syberg 2007, 213). The trick was to select scions that could promote fruit varieties suited to the Nordic region (Olafsen 1898). Through grafting that combined compatible stems and shoots from here and there, the Norwegian fruit frontier came to life, breaking new ground for Nordic cultivation and doing so by way of imported natural parts.

A closer look at the practice of grafting might help us think about the pioneering work it took to make a Norwegian apple domus as the fruit was moved north and planted in new fertile ground. An apple seed is genetically a mix of the mother tree that grows the fruit and an unknown pollinator variety. This means that every apple carries the seed of a new variety (*Pometet*, University of Copenhagen). Planting a seed, then, is no guarantee of reproducing the qualities of the apple that once held it. Grafting is thus the way to imitate nature through attempts at controlling a relation between two known trees, by bringing them in touch with one another and literally sealing their alliance (with beeswax, most often). The resulting tree population is ideally uniform and continuous; apple trees left to nature's vagaries may run astray, which is not the way to make fruit production commercially profitable. What interests me here, in terms of discussing how the Norwegian apple domus gets coordinated on the northern fruit frontier, is the deliberate inclusion of foreign fruit components over time. At some point in the story of apple cultivation in Hardanger, it seems, the grafted apples' combination of varieties from elsewhere was perfected enough for the resulting apples to become indigenous. In Lofthus, at least, near the Bioforsk research center and in other fruit villages around the fjord, I come across signs explaining that a given orchard participates in restoring "Hardanger's natural heritage," which implies that genetic resources for all the original Nordic fruit varieties should be preserved as a natural bank account of the region before they are lost. Through centuries of coordination, the Hardanger apples have become what they are and found a home that now needs to be protected.

Around the year 1900, producers of apples and other fruits (pears, plums, and sweet cherries) in the area began applying fertilizer systematically, resulting in a steep increase of the amount of fruit produced. Fertilizing stimulants were welcome new arrivals in the Hardanger apple homeland. The Ullensvang Horticultural Society was founded in 1897 with the aim of coordinating orchard development and commercial potential, and by the early twentieth century the local fruit producers had started covering new land, not least thanks to the new and improved possibilities of using the fjords' waterways as transport infrastructure, connecting producers to fruit exhibitions and sales events across the country and internationally and enabling them to reach emergent markets (Syberg 2007, 214 ff.; Austad and Hauge 2008). Again, the apples and other products from the fjord area broke new ground and gained wide acclaim wherever they went—eventually so much so that the Hardanger label was adopted as a trademark to both express and affirm the unique quality of the local produce. The new and intensified connections to other places and products also made the Norwegian fruit newly distinct.

Little by little, cooperatively owned cooling, packaging, and storage facilities—predecessors of the organizations now behind the *Dyrk Smart* campaign—popped up, gradually systematizing the export of Hardanger fruit to places near and far (Syberg 2008). In the first half of the twentieth century, the fruit industry grew in Hardanger, as elsewhere, owing to new technologies that made it possible to store fresh fruit for longer periods (Freidberg 2009).

This long story of fruit cultivation in the Hardanger region may seem like a smooth tale of progressive commercialization and industrialization, a gradually increasing mastery of nature by people with business interests at heart who continuously worked to consolidate what medieval clergy had started. However, industrial progress was seemingly a mixed blessing for the fruit producers in the Western fjord region; it posed a threat to the domus even while supporting it. In volume 2 of Cappelen's reference work *Norge* ("Norway"), the *Geographical Encyclopedia,* from 1963, which I leaf through outside a secondhand shop in Lofthus, I read that fertilization and spraying are hard work for the fruit producers, who must invest in costly technology to control nature's vagaries: "Every year they have to make great effort and obtain knowledge about fertilization and spraying and invest large amounts of money in pumps, machines and other technical equipment in order to prevent damages that are easily caused by wet, frost, cool temperatures, wind, insects and other evils" (1963, 131).[3]

Judging from the encyclopedia and its postwar ring, by the middle of the twentieth century, domestication as the tough effort to beat the hostile Norwegian nature by means of pesticides and machines seemed to be the way to go. Even if these new participants in the making of the apple homeland are costly, pumps and other technology can help fight various evils. But this story of progress is not smooth; temptations to leave the fruit trade lurk, and industry, for all its dirty emissions, is an alluring modern adventure. Likewise, the increased competition characterizing the increasingly international market calls for counteraction. Consider this, rather long, quote from the 1963 encyclopedia:

> But fruit from the Western region is nowadays met with competition both from villages on the plains further east and southern countries. From the latter competition is not likely to decrease in the future. An important counter measure is first class cooperative storages that can offer Norwegian fruit well into the spring. The farmers certainly have a full day's work all year round in their orchards and there have already been many threats and temptations to leave this vertical life mode. By the southern fjord, Hardanger, in Kvam, and the inner parts of Sogn foul smokestacks rise to the sky from the industry. In any case, in Hardanger many people have preferred to stick to their fruit. But it cannot be denied that the industry is a modern adventure in a positive sense for the villagers by the fjords and will benefit the whole country.[4] (1963, 132)

What we see here, I think, is a bumpy story of pioneer work on the Norwegian fruit frontier. Industrialization in the guise of new technologies is both a danger to the region, tempting people to give up the demanding "vertical life mode" on the orchard slopes and leave the apple domus for more adventurous livelihoods, and a countermeasure abetting the local fruit trade by using state-of-the-art storage facilities to curtail competition from imported goods and thus fortify the local apple land. Long before the *Dyrk Smart* campaign, sticking by one's fruit, however, is what many people in Hardanger preferred in any case.

What I find fascinating about the quotes from the encyclopedia and the fieldwork conversations presented above is the ambiguity they embody as to what it takes to make a livable apple domus on the fruit frontier. What coordination is needed to continue and perhaps strengthen the local fruit production? Do producers need completely new technological measures? Or do they just need to stick by the fruit? And what about the foul industrial smoke? Might continuity be ensured by practices like grafting—a freezing of sorts that makes perfect copies of the fruit that is already rooted there or that time has shown to be suitable for inclusion in the Nordic natural heritage? Or is cooperative organization the way forward?

In a video on YouTube produced as part of the *Dyrk Smart* campaign, the voice-over matter-of-factly says that "the experiences of the previous generations are no longer relevant," referring, I suppose, to the full day's work year-round also mentioned above and to the traditionally meager profit. With new methods for apple production, the video explains, it simply pays to enter the trade and move into the Hardanger apple domus. The area is indeed ripe with easily picked fruit. Bioforsk, enacting Norwegian state-sanctioned ambitions, promotes organic production, which is also mentioned in the *Dyrk Smart* material. However, so far few of the Hardanger producers grow organic fruit, although generous subsidy schemes are in place for producers who want to replace conventional methods of fruit production (Nes 2014). According to Eivind from Bioforsk, this skepticism toward organic production has to do with the long history of cultivated fruit in the region. The story among the producers, Eivind tells me, is that there is no need to fix what is not broken, namely, successful production in operation since the monasteries tended the orchards in the area. This, it would seem, is regarded as a natural and cultural heritage that is protected through sustained practice; local fruit producers, on the other hand, refer to organic farming as merely "a philosophy." Eivind and I discuss this invocation of historical legitimacy as a curious argument, given that the monks' version of conventional farming is surely very different from what goes on today, even granted that producers eight hundred years ago did not label their practices "organic." Part of Hardanger fruit production is to carry on performing the apple homeland, it seems, and to continue what the monks started, even if it means a steady application of pesticides to expel unwelcome inhabitants. If you want to survive and indeed profit as a Hardanger fruit producer, you cannot leave nature to nature, even in areas with perfect conditions for production.

What I have intended to show here is that, curiously, grafting, fertilizing, implanting, nationalizing, preserving, and pesticide-spraying, among other practices, play into naturalizing the Norwegian apples on the slopes around the Hardangerfjord. This performs a complex kind of frontier work that coordinates the domus so that it appears natural and remains livable. The pioneering work of producing Norwegian apples as specific natural goods is a means of relating to other foreign apples, nationalized to take root in Hardanger as natural heritage, to a tempting and yet repulsive industry, and to organic farming as an unworldly philosophy, a claim backed by assertions that clergy and farmers in the Middle Ages did not buy into such fads. Through a coordination of these connections, among many others, in the fjord region, actors involved in Norwegian apple production make it possible to continue the fruit craft on the vertical Western fruit frontier where apples and other fruits are naturally at home.

Cutting Associations: Interventions for the Win

Around the year 2000, according to Eivind from Bioforsk, the Norwegian fruit market suffered from a virtual collapse. The large retailers and supermarket chains pushed prices down by pointing to cheaper imported fruit from abroad. In recent years, however, cooperative organization has helped the Hardanger producers take back some of the power and collectively indicate the reward of local fruit production. The producers and their cooperatives know that they provide goods that are in demand, and increasingly so in light of trends favoring locally produced goods and high quality food (cf. Pratt and Luetchford 2014). When I visit Ullensvang Fruktlager (Ullensvang Fruit Storage) in Lofthus, the manager of the cooperative explains: "We could easily double or even triple our orchards' yield without satiating the market—if only we had the acreage available to do so. We can't really catch up with the demand for local fruit, because the mountains around the fjords leave little room for expanding the orchards." Norwegian apples account for about 20–25 percent of the total consumption in the country, so there are more market shares to be conquered. However, instead of expanding the orchards, which is impossible because of the steep incline of the mountains, the fruit storage manager explains, producers need to increase the yield per hectare. This is one of the purposes of the *Dyrk Smart* project—to promote methods of production that are more efficient, and thus economically attractive, involving more closely planted and slimmer trees with fewer leaves to shade the fruit from sunlight, as well as more plant fertilizer and built structures to support branches near the ground. Thinning the apple trees is the key to preventing branches from growing wild and wasting precious arable land, the *Dyrk Smart* campaign instructs.

Ullensvang Fruit Storage is prepared for the increased efficiency that is supposed to make the Norwegian apple domus competitive. The facility seems to have jumped the gun of the race to produce ever larger amounts of apples. Ready to accommodate an insatiable market for local fruit, the storage facility has expanded within the past year. The manager shows me around and points out parking lots, storage buildings, and cooling rooms that are all newly constructed and ready to be filled, in anticipation of this year's yield and of growing harvests to come. As we saw from the *Geographical Encyclopedia*, efforts to prolong the season—or more accurately, the period during which consumers can buy Norwegian products—have long been considered a factor in the competition with imported goods. Many fresh fruits are difficult to keep at their best for long; the ripening tends to go on after picking. As a consequence, experiments in refrigeration combined with controlled atmosphere storage have evolved along with the industrialization and commercialization of fresh fruit. Controlled atmosphere technology, first tested in the late 1940s, entails low temperatures and oxygen levels in storage rooms to arrest ripening (Freidberg 2009, 125). At Ullensvang Fruit Storage, the manager describes state-of-the-art versions of this controlled atmosphere practice, developed in close collaboration with the Bioforsk unit in the village, while we pass the closed doors of sealed-off cooling rooms with carefully designed contents of air, soon to become sanctuaries of Hardanger fruit.

Other interventions are at play in attempts to control the atmosphere, so to speak, of local apple production, organizing relations so that yet more distinctions appear on the Norwegian fruit frontier. When I meet with Johan, one of Lofthus's senior producers, in his orchard, he and a crew of young farmhands from Poland are thinning the trees, removing leaves and small apples. He tells me about the *vernetoll* (protective toll), a tax applied to imported apples when Norwegian apples are in season, mid-August through November. This toll raises the price of the otherwise cheaper imported apples from South America, New Zealand, or southern Europe, enhancing the Norwegian producers' ability to compete. The fencing around the Hardanger apple domus is built partly by state policies. Half-jokingly, Johan asserts that, of course, he and his neighbors produce far better apples than the imported ones, so with regard to quality the Lofthus orchards have no need for the kind of affirmative action implied in the toll. Tongue-in-cheek, he tells me that, in fact, it only helps him build more evidence of the superior quality of his apples, because the toll causes more people to buy Hardanger fruit, if the Argentinian apples are suddenly almost as expensive. The vernetoll merely makes it easier for Johan to prove his point—with the aid of a state that buys competitors out of the market, making their entrance into the Norwegian homeland more difficult.

At Bioforsk a few days earlier, Eivind talks about the toll and explains it in terms of both political will to keep the Hardanger region afloat as a fruit producing area within the national resource landscape of Norway and as means of accommodating skeptical local consumers who are not convinced that foreign products are sufficiently clean, even if they have passed all sorts of tests that check for leftover pesticides and other chemicals. Eivind tells me that even though the imported goods comply with standards of allowed limit values of chemicals, many Norwegians still trust the local products to be cleaner, even if they are produced conventionally. Interestingly, the pesticides that I learned about in the *Geographical Encyclopedia* as a means to enforce the Norwegian apple homeland seem to have been naturalized in present-day consumers' eyes, setting these chemicals apart from the foreign pesticides that might be found in imported apples. Again, the fruit frontier work operates to coordinate relations to apples from elsewhere by distinguishing between local and imported chemicals and the prices these different apples can fetch.

Loud pop music meets me in one of the uppermost orchards of Lofthus. Walking closer, I trace the sound to a small truck parked on a gravel road between the apple trees. Three men speaking no Norwegian and very little English look back at me. We manage some conversation, through which I learn that they are from Poland and are employed as seasonal workers for about four months. Right now they are cutting leaves from the trees—"for the right color," the youngest guy tells me, indicating that sunlight is what causes the apples to develop the right hue on their skin. "It is good work," they tell me, when I ask them what they think of the place. During her break, I run into Bianca, a Bosnian girl who has come to Lofthus for nine seasons to work as a fruit packer at Ullensvang Fruit Storage. Pointing to her cigarette and coffee, she laughingly tells me that she is trying to replace both of these with apples. When she first came north to work, she did not much care exactly what she was packing, but the apples seem to have grown on her. She now studies ecology back home and funds her studies by her Norwegian income. The manager of the Ullensvang Fruit Storage tells me that this year he has had to call in workers like Bianca almost a month earlier than expected due to an unusually warm and sunny summer, which has caused the season to begin at least two or three weeks earlier than usual. Luckily, over the years he has formed a crew of reliable guest workers who keep coming back. The local youths, he says, are not interested in seasonal work in the fruit trade, and the Hardanger orchards thus depend on people who are willing to work for what Norwegians apparently consider to be cut wages; the borders of Europe are apparently both present and permeable enough to make the migrant workers come and contribute to producing Hardanger fruit.

My point here is to show how the Norwegian apple domus is coordinated through a series of cuts. Impeding the smooth influx of fruit from a global market combines with attempts to arrest the ripening of apples through the cool removal of oxygen. Attracting border crossers in the guise of experienced migrants, but not chemicals, goes hand in hand with thinning trees to optimize the yield per hectare. A deliberate cutting off of too-close associations with other apples makes the Norwegian fruit frontier appear as a pioneering site of world-class production.

Norwegian Apples in Controlled Atmospheres?

"It has been proposed that the apples grown here in older times were wild forest apples. This is impossible. From the 13th century, expressions in the old laws . . . refer beyond doubt to cultivated apples; otherwise one would not have used the term 'Apple orchard' [Eplegard]. A different question, however, is where these apple trees came from. Were they wild trees that had moved in, and which later by repeated seeding developed into refined sorts? This is not likely to have been the case. The most probable is to assume that trees, scions and seeds were brought from other countries; monks and monasteries were the brokers" (Olafsen 1898, 25).[5]

Cultivation of the *Malus domestica* in Norway goes back a long time and builds on a range of performative practices (cf. Lien and Law 2011). Medieval laws specify rules for apple orchards, as the fruit cultivation traveled north as a clerical chore. Moving north is obviously a relative shift, and what I have meant to show here is that the fruit frontier thus created keeps enfolding its relations to the south. Think only of the *Dyrk Smart* vocabulary arguing that the northernmost apples are the best *in the world*. My point here is that the work to coordinate the Norwegian apple domus as a unique production site is what makes the Hardanger apples worldly. The relational work on the part of the smart pioneers on the fruit frontier performs a paradox: the more the Norwegian apples are and have been rooted, implanted, refined, singled out, protected, and cultivated as the best in the world, the more they come to be in touch with that world. In other words, the forging of connections and the cutting of associations between Hardanger apples and their "other" come together in the creation of the northern fruit frontier and its perfect products. Whether the apples are grafted, as it were, onto a long story of continued local cultivation by way of experiments with and refinement of foreign apple components cultivated elsewhere or are distinguished as specific and unique local resources through toll borders and high-tech storage, the fruit frontier in Hardanger comes to life as an arena for implicit comparison and connection with potentially all other apples in the world, thereby enacting a specific natural good. The taste of place is carefully crafted (Paxson 2010). The pioneering figures working this frontier are out on a complex errand to produce perfect(ed) Hardanger fruit; on the one hand, they stress the continuity of apple production in time and space and, on the other, they engage in all kinds of practices to cut connections to other apple locales and stories.

This is not just an argument about the fundamental integration of global currents, markets, and supply chains in our time, though it is that too (cf. Pratt and Luetchford 2014; Tsing 2009). Taking my cue from the apple producers, researchers in Lofthus, the gardening enthusiast Olafsen, and the *Dyrk Smart* campaigners, I also mean to see the Norwegian apples as a resource through which to explore the making of a particular apple homeland—in this case a frontier landscape where fruit quality is seen as both naturally rooted and affirmatively protected.

In what ways does this composite fruit frontier work shed light on domestication on the margins? For one thing, it seems to me to qualify what a margin is, namely, a site that thrives through being both in touch and cut off from what it demarcates; the Norwegian apple domus cannot but work as a part of a single apple world. Further, I think that the promoters of Norwegian apples embody a basic conundrum in resource ecologies of our time: the ambition to produce resources differently as specialist products is tied up with the ambition to succeed in an integrated cutthroat market. The more the Hardanger fruit cultivation is stressed as an alternative to and better system of fruit provision than the heavily commercialized supply chain, the more it must buy into the mainstream methods of efficient production. The question for producers is how to gain a competitive edge through the distinction of local quality and storytelling while surviving in a time-space compressed market, where such distinctions do not always matter and/or where they need investments of time and money to stand out as important.

This question is also relevant to frontier work as a form of domestication; the new ground beyond the frontier and the well-trodden fields inside it are mutually constitutive. The *Dyrk Smart* campaigners and the people I met during my fieldwork thus seem to perform a very circumscribed form of apple activism, as they try to work the frontier both to create more natural and cleaner products, saving the world and the future, as we read in the campaign flyer, and to optimize production and marketization through fertilizing, thinning techniques, subsidies, national borders, and so on. The fruit frontier implodes, one might say. If the motor for the prototypical tropical plantations so integral to colonial expansion in the nineteenth century was annexation of new grounds (cf. Tsing 2012), here the frontier work is a kind of affirmative intensification of production on a limited range of Nordic land, set apart from but completely entwined in a globalized fruit market.

Accordingly, and this is perhaps where I see the critical potential in this kind of resource anthropology, it seems to me that a position outside the "capitalocene" landscape, to use Haraway's term, is very difficult to locate, and thus all the more important to explore. If, indeed, we inhabit one world of continuous becoming-with others, for better or for worse, new forms of collaboration and dialogue are needed to inhabit resource frontiers. Exploring an apple domus, then, is not necessarily to locate a master narrative of increased command but to engage in an ongoing and analogue conversation with different landscapes, histories, kinds of material, interests, and perspectives to pay attention to what is allowed in a domain and what is not (cf. Hastrup 2014). The Norwegian apples become frontier resources in more senses than one, as they work as a genuinely relational method for thinking about fruit, places, markets, and world—experiments in controlled atmospheres, indeed.

NOTES

1. All translations of written material from Norwegian and Danish are by the author. The original quotes are given in the footnotes.

2. "Som det sees af det ovenfor anførte, antager vore Historieforskere, at Frugtdyrkning og Havebrug fra Udlandet, d. e. England, Danmark, Tyskland og Frankrig m. fl. Lande har udbredt sig til de norske Klostre, og at disse Fredens stille Sysler har fundet en frugtbar Jordbund inden Klostermurene" (Olafsen 1898, 8).

3. "Hvert år må de legge ned en masse strev og lærdom i gjødsling og sprøyting og ha store udleg i pumper, maskiner og andet teknisk udstyr for at avverge de skader som let kan forvoldes af væte, frost, kølighed, vind, insekter og annen styggedom" (Olafsen 1898, 8).

4. "Men vestlandsfrukten ellers møter nå til dags konkurranse både fra slettebygdene østpå og fra det sørlige utland. Det sidste bliver vel ikke mindre etter hvert. Et viktigt mottrekk er førsteklasses felleslagre, som kan tilby norsk frukt til langt ud på våren. Fruktdyrkerne har sannelig hel dagsjobb hele året med sin frukthage, og trusler mod og lokkemidler bort fra dette vertikale tilværet har der været mange av allerede. Utetter Sørfjorder, Hardanger, i Kvam og innerst i Sogn stiger dett fæl royk mod himlen fra industrien. I Hardanger i hvert fald har folk i stor udstrækning foretrukket at bli ved sin frukt. Men det kan jo ikke nektes at industrien er det moderne eventyr

i positiv forstand for bygdene derinne, og med gode virkninger for det ganske land" (Olafsen 1898, 8).

5. "Der har været fremsat den Formodning, at de Æbler, som dyrkedes hos os i den ældre Tid, skulde være vilde skovæbler. Dette er umuligt. Udtrykkene i de gamle Love viser . . . at der fra det 13de Aarhundrede utvivlsomt menes dyrkede Æbler; ellers kunde der ikke bruges Udtrykket Eplagard. Et andet Spørgsmaal er imidlertid, hvor man fik disse Æbletræer fra. Er det vilde Træer, som var flyttede ind, og senere ved gjentagende Frøudsæd havde udviklet sig i ædel Retning? Det er ikke tænkeligt, at saa har været Tilfældet. Det rimeligste er at antage, at man fik Træer, Kviste og Kjerner fra andre Lande; det var Munkene og Klostrene, som her var Mellemmænd" (Olafsen 1898, 25).

REFERENCES

Abram, Simone, and Marianne Elisabeth Lien. 2011. "Performing Nature at World's End." *Ethnos: Journal of Anthropology* 76, no. 1: 3–18.

Åsen, Per Arvid. 2009. "Plants of Possible Monastic Origin, Growing in the Past or Present, at Medieval Monastery Grounds in Norway." In *Plants and Culture: Seeds of the Cultural Heritage of Europe*, ed. Jean-Paul Morel and Anna Maria Mercuri, 227–38. Puglia: Edipuglia.

Freidberg, Susan. 2009. *Fresh: A Perishable History*. Cambridge, MA: Harvard University Press.

Haraway, Donna. 2008. *When Species Meet*. Minneapolis: University of Minnesota Press.

Hastrup, Frida. 2014. "Analogue Analysis: Ethnography as Inventive Conversation." *Ethnologia Europaea* 44, no. 2: 48–60.

Lien, Marianne, and John Law. 2011. "'Emergent Aliens': On Salmon, Nature, and Their Enactment." *Ethnos: Journal of Anthropology* 76, no. 1: 65–87.

Nes, Jan Ove. 2014. Økologisk dyrking av frukt*: Er dette noko for deg?* Accessed April 14, 2015. http://www.dyrksmart.no/assets/filer/Notat-kologisk-dyrking-av-frukt-v02.pdf.

Norge. 1963. *Geographical Encyclopedia, Volume II*. Oslo: J. W. Cappelen's Forlag.

Olafsen, Olafsen. 1898. *Frugtavl og Havebrug i Norge i Middelalderen*. Christiania: Grøndahl og Søns bogtrykkeri.

Paxson, Heather. 2010. "Locating Value in Artisan Cheese: Reverse Engineering *Terroir* for New-World Landscapes." *American Anthropologist* 112, no. 3: 444–57.

Pometet. Danmarks genbank for frugt og bær. Handout on grafting by the Pometum, Department of Plant and Environmental Science, University of Copenhagen. Acquired in 2015.

Pratt, Jeff, and Pete Luetchford. 2014. *Food for Change: The Politics and Values of Social Movements*. London: Pluto.

Syberg, Karen. 2007. *Æblets fortælling*. Copenhagen: People's Press.

Tsing, Anna. 2005. *Friction: An Ethnography of Global Connection*. Princeton: Princeton University Press.
Tsing, Anna. 2009. Supply Chains and the Human Condition. *Rethinking Marxism: A Journal of Economics, Culture and Society* 21, no. 2: 148–76.
Tsing, Anna. 2012. "On Nonscalability: The Living World Is not Amenable to Precision-Nested Scales." *Common Knowledge* 18, no. 3: 505–24.

8

DOMESTICATION OF AIR, SCENT, AND DISEASE · *Rune Flikke*

How can air be domesticated? Domestication is most often thought of as a kind of interspecies relation in which humans modify animals or plants, bringing them under their control and shaping them into providers of food, protection, and labor. But might what initially seem to be disembodied—or even "empty"—spaces also be sites of domestication projects?

In this chapter, I will introduce a case study from King William's Town on the Eastern Cape in the mid-1870s, where transplants of antipodean eucalyptus trees emerged as an attractive means through which the settlers could shape their environment and their lives. Eucalypts are used in plantation forestry and have been subject since the early nineteenth century to breeding and hybridization practices that could certainly be labeled "domestication." Yet in this chapter I am not primarily interested in the domestication of eucalypts themselves but will focus on eucalyptus to investigate human relations to air and atmosphere in Victorian South Africa. Instead, the analytical focus will be on the olfactory traces they left in the atmosphere. Air, I suggest, became a sight of domestication, because it was conceived as containing material aspects of a wild, threatening, and contagious "otherness" that needed to be controlled.

This connection becomes apparent when we consider that colonial South Africa was guided by Galenic medicine and the notion of miasma as disease-bringing air caused by putrefying organic materials (cf. Pelling 1978; Sargent 1982). Miasma as the source of disease opened what David Armstrong (1993) has referred to as a "critical public health space" between humans and landscapes. I will argue that the settler communities planted eucalypts and invested time and resources in scientifically altering trees so they would thrive in the South African soil in order to tame the threatening and poisonous colonial atmosphere (cf. Flikke 2016a). This occurred at a time when the inhabitants of King William's Town experienced a decimation of its population due to flu, smallpox, and typhoid in addition to lung-sickness and other epizootics that threatened their livelihood (e.g., Peires 1989, 71).[1] Other studies have shown how epidemics spread through colonial Africa along trade and migrant networks at a rate and with a severity that gave rise to the trope "the white man's grave," hence embedding the particular history of King William's Town in a larger history of the colonial conquest (Feierman and Janzen 1992; Comaroff 1993).

I will argue that air, as the medium through which we live, is a contact zone that wipes out any clear distinction between subject and object. This was accentuated by the Galenic understanding of the disease vector through which air challenges the integrity of the human body and its boundaries. Breathing meant inhaling and merging with a threatening otherness, and planting eucalypts was a way to domesticate the disease-bringing atmosphere. The domestication of air hence only makes sense when we view domestication as a project of self-creation—a project intimately involved with the maintenance, and reshaping, of body boundaries in a colonial world where the integrity of the subject was constantly threatened by the traces of African otherness that were fused with the air.

Recently, air has increasingly been the focus in anthropological studies (cf. Choy 2011, 2012; Choy and Zee 2015; Ingold 2010, 2011; Zee 2015). However, despite the fact that medical historians have emphasized the connections between the tropical atmosphere and miasma (cf. Arnold 1996, 5–10; Curtin 1989), air has received little analytical attention in the context of the colonial conquest (cf. Flikke 2016a). I will, however, start with Sloterdijk, who recently defined the twentieth century by its relation to the atmosphere.

Atmosphere as Aerial Domus

Sloterdijk (2009) argues that the twentieth century was marked by three things: i) the practice of terrorism, ii) the concept of product design, and iii) environmental thinking. Based on these criteria, Sloterdijk suggests that the twentieth century started on April 22, 1915—the date when the German "gas regiment" launched their first attack with chlorine gas in combat, bringing the three traits together in one expression. Rather than the enemies' bodies, their environment was now targeted, undermining its life-sustaining ability.

Caught in such a battle over the environment's ability to provide sustenance, humankind could no longer take the atmosphere for granted. Sloterdijk argues: "Those areas affording inhabitants and travelers in the air milieu a natural rapport to the atmosphere and an unquestionably given and anxiety-free, unproblematic being [were lost, thus giving the Europeans] the decisive push into modernization" (2009, 47).

Though I find the larger tenets of Sloterdijk's argument both interesting and important, it is too Eurocentric to make sense of the colonial situation. First, the burgeoning literature on climate change has pointed out that the atmosphere cross-culturally has tended to be viewed as the sphere of the gods and the divine, not as given and anxiety-free (Donner 2007). In many contexts, then, a fluctuating and unruly atmosphere tends to surface as a source of uncertainty that needs to be ritually regulated and "domesticated" to create a needed sense of order and predictability—a dwelling place, a domus, in the colonial periphery. Mike Hulme (2014) has suggested that climate, understood as atmospheric conditions typical of a given region, is one such metaphor that embeds the atmosphere in human projects that create a sense of stability and consistency, carving out spaces for meaningful human action and interaction. Second, this was the case in precolonial South Africa, where the environment was viewed as a generator of the poisonous gas miasma, which turned the air the settlers were breathing into a source of uncertainty.

As Sloterdijk is aware, his concern can be traced back to *The Arcades Project*, wherein Walter Benjamin argues that the nineteenth century, "like no other century was addicted to dwelling" (1999, 220). This addiction was marked by an irresistible drive to construct a casing, insulating certain spaces for living from the adjacent atmosphere. This did not take the same form in the colonies as it did in the colonial centers such as Paris and London, where by the mid-nineteenth century glass-roofed arcades bound blocks of buildings together, letting light through and increasing visibility as it enabled people to see and be seen while sheltering them from the elements (cf. Benjamin 1999, 873). Though these steel and glass-covered structures were absent in the colonial periphery, the need to control the air opened other forms of "air conditioning," understood as efforts to "disconnect a defined volume of space from the surrounding air" (Sloterdijk 2009, 20). I will argue that the settlers inhabited a world where they tried to control the air they were breathing through reshaping the landscape and planting eucalyptus so that "such offensive matters could be deodorized."[2]

A quick overview of the Victorian acclimatization societies will underscore how closely the activities of these societies were tied to efforts not only to establish homely, ambient, and properly "domesticated" spaces in a foreign land by re-creating aspects of the landscape left behind but also to secure healthy and dependable places to dwell in the uncertain colonial atmosphere.

Acclimatization Societies: Colonialism as Domestication

Currently, the global transplant of species is a highly contested and politically potent topic. The recent literature on the global flows of plants and animals have primarily focused on material exchanges and their consequences, such as the notoriously unstable boundaries between alien and invasive species and how the visual landscape has not only changed but also could be read as an indicator of larger political challenges related to national character and integrity (e.g., Bennett 2014; Comaroff and Comaroff 2001; Lien 2007). As a quick look at the Victorian acclimatization societies reveals, alien species have not always been viewed as a problem.

Studies on the growth of acclimatization societies (e.g., Lever 1992; Osborne 2001) have pointed out that even though species had privately been diffused for a long time, interest in acclimatization as a scientifically mediated process of transplanting usable organisms to increase productivity and profit dramatically expanded in the mid-nineteenth century. Though experimental transplants of species have a long history, the first acclimatization societies were established in Paris, in 1856, and London, in 1860, and quickly spread to the colonial peripheries. Despite differences between the French and the British approaches and multiple and sometimes conflicting concerns, the societies were largely occupied with three things. First, their growth was closely related to efforts to domesticate the colonies through the spread of botanical knowledge and science to cultivate and exploit the new territories and develop the colonial as well as domestic agriculture (cf. Brockway 2002; Lever 1992, chapter 3). Second, European species were also spread for the "practical and immediate utility to the country gentleman" (Burgess 1968, in Osborn 2001), thus promoting English sports and gamesmanship, for instance, by stocking suitable waters with trout and salmon. This is related to my final point: the acclimatization societies were an integral part of "the settler's continuing attempts to come to terms with their new lands, to find their place in the country and its place in them" (Dunlap 1997, 304), creating an environment reminiscent of the land they left behind. This required transforming the colonial landscapes according to ideals of the homeland (e.g., Lien and Davison 2010). The close connection between the acclimatization societies and the cultivation of new territories and domestication of new species is further visible in the slippery terminology of the time, wherein "acclimatization" was used interchangeably with "naturalization" and "domestication" (cf. Osborne 2001, 137). For example, in 1860—the year the Acclimatisation Society of the United Kingdom was established—the *Saturday Review* magazine defined "acclimatization of animals . . . [as] the reduction into a state of domestication of wild species" (Lever 1992, vii). Hence, I suggest that the activities promoted by the acclimatization societies need to be accounted for within a theoretical framework that treats them not only as aesthetic efforts to re-create familiar environments but also as charged existential projects that aimed to tame uncertainty and unpredictability in a precarious situation in a foreign climate with hot, humid, and disease-bringing air.

The important role played by the acclimatization societies' scientific approach to transplanting usable species becomes apparent when we look closer at the particular case of the massive global expansion of eucalypts.

Domesticating Eucalypts at Home and Abroad

Eucalypts are evergreen hardwood trees that belong to the myrtle family (*Myrtaceae*), and they are part of a botanical group that consist of more than seven hundred species. With the exception of fifteen eucalypts that appear naturally in New Guinea and Indonesia, all are native to Australia and Tasmania and evolved in adaption to the acidic and low-nutrient Australian soil. As an intentionally introduced species, eucalypts are a rarity, as they moved from the periphery to the colonial centers, though several researchers have since challenged this idea, pointing out that plants do not necessarily "follow the flow of power" (Beinart and Middleton 2004, 10; cf. Bennett 2011, 127f.). In his seminal study on global environmental transformation, *Ecological Imperialism* (1986), Alfred Crosby argued that species transfers mostly involved a one-way dispersal of Old World species throughout the New World. Crosby proposed that this unevenness of flows could, in part, be explained by the evolutionary benefits inherent to the large east-west span of Eurasia (cf. Diamond 1997, 183–86). This is important since many plants and cold-blooded animals have no, or poor, inbuilt temperature control and will be severely affected by movements outside of their natural latitudinal range, at the same time as the variations in soil compositions, altitudes, and precipitation within a climatic zone is an added evolutionary bonus.

Crosby later acknowledged that the eucalypts were a rare example of a New World genus that, due to its biological ability to flourish under highly variable conditions, successfully colonized Europe and the Americas (Crosby 1993, xiv). However, Brett M. Bennett has convincingly argued that the global success of eucalyptus was less a product of their innate adaptability than of explicit human efforts to transplant them. Bennett's work suggests that eucalypts did not initially thrive in South Africa when they were introduced in the early parts of the nineteenth century. Part of the challenge was that the eucalypts that eventually took root in South Africa were often hybrids with very different qualities from those reported from the antipodes (Bennett 2011, 131). Hence the timber lost a number of its economically valued traits that contributed to the dissemination of the species in the first place (cf. Sim 1905, 158). Furthermore, the significance of the acclimatization societies indicate that the global expansion of eucalypts succeeded not due to "natural advantage" but due to a century of testing and scientific experimentation with species selection and the creation of new hybrids (cf. Bennett 2010, 29).

The scientific work to acclimatize eucalypts to the ecological conditions of other continents is surprising given that historians have documented that the early settlers in Australia "hated [Australian] trees" (cf. Bennett 2010, 125f.; Bonyhady 2000, 178–82). Eucalypts were described as gigantic and useless "weeds" (Bonyhady 2000, 181) or were simply reduced to "containing so many tons of firewood" (Bonyhady 2000, 183). There were several reasons that these negative attitudes dominated the local species in the antipodes, of which I will outline three: First, the trees were hardwood and a considerable obstacle to the dominating pastoral economy, since they were hard to chop, making the clearing of new pastures very labor intensive. Second, they were evergreen trees that early on were considered anomalies, since they typically shed their bark and not their leaves, which are covered with oil glands and tend to hang vertically on mature trees, thereby providing patchy shade that rendered them unappealing to European aesthetics. Hirschfeld can help us shed more light on these attitudes. In his five-volume work on garden art published between 1779 and 1785, Hirschfeld argued how, by the eighteenth century's end, the English garden was already a careful orchestration in which a sophisticated, contemplative, and "softly melancholic region occurs in the absence of vistas; through depths and low areas, dense thickets and woodlands, often simply through groups of tall, closely spaced trees with thick foliage . . . low-hanging, dark, or dusky green leaves and deep shadows spreading everywhere" (Hirschfeld 2001, 187f.). Within such an aesthetic field, the eucalypts, with their thin canopies of drooping evergreen leaves, incapable of providing proper shade and always revealing the contours of the distant landscape, were unable to provide a serene and contemplative ambiance (cf. Lien 2007). Australian landscape alterations thus largely focused on ring-barking eucalyptus, replacing them with imported European trees such as oaks and pines to "improve" the visual landscape by transforming it according to the sensibilities of a European landscape aesthetic, helping the settlers "to find their place in the country and its place in them" (Dunlap 1997, 304).

Finally, Adrian Franklin has pointed out how eucalypts as a species are characterized as a "specialist in exploiting disturbance," which through "a dance of agency" gradually replaced the rain forest through natural ignition over the millenniums (Franklin 2006, 562) to the point of becoming so resistant to fires that they depend on them for reproduction (Hay 2002, 210ff.; Pyne 1992). The eucalypts were hence associated with fires, which in minutes could consume what had taken decades to build.

However, these negative antipodean voices were to a certain extent contested (Bonyhady 2000, chapter 6). Ferdinand von Mueller, also called Baron Blue Gum, was a major force in shifting the initial flow of species from one of colonial centers to the peripheries, which were typically assumed to contain inferior species, to one of the peripheries to colonial centers (cf. Hay 2002). Having traveled to Australia from Germany at the age of twenty-two, von Mueller followed the lead of French scientists who had been the primary promoters of eucalypts as therapeutic agents (Bonyhady 2000, 171). He was an ardent proponent of the transplant of eucalyptus trees, which he viewed as an "antimalarial agent," from the early 1850s until his death in 1896 (e.g., Bennett 2011, 135). He listed more than 1,500 publications on eucalypts, collected and cataloged 90,000 samples, and mailed seeds by the thousands to South Africa and to other sites across the globe (Hay 2002; cf. Osborne 2001, 148f.; Sim 1905, 150).

Though these negative antipodean attitudes began to change toward the end of the nineteenth century, it was only toward the end of the twentieth century that eucalypts became a regular part of Australian timber plantations (Bennett 2011, 138). Overall, colonial settlers in Australia repeatedly saw eucalypts as foreign and distasteful; these were not trees that in any way made the continent seem "homey" or more habitable. The paradox is that this occurred at the same time eucalypts were spread around the globe, fueling a wanting economy (cf. Flikke 2016a) and slowly changing global ecologies as they found new homes in Africa, Asia, Mediterranean Europe, and North and South America (Zacharin 1978). Both the local eradication and the global embrace of eucalypts were directly linked to the particularities of the genus. As I have argued elsewhere, the global transplant was not only for economic and landscape-modifying purposes but also as a means of purifying the atmosphere, ridding it of miasma through its hygienic smell (Flikke 2016a; cf. Bennett 2011, 127). Though eucalypts were considered an aesthetically inferior species that created a sense of estrangement, during epidemics and health scares its oil glands and olfactory presence transformed it into an active producer of healthy atmospheres—it purified the air.

The eucalyptus oil contained in the leaves and bark was not merely flammable; eucalypt oils are also known for their antiseptic qualities and are used in a wide range of products such as soaps, industrial solvents, perfumes, and foods and are widely recognized as a health product for relieving asthma and other respiratory problems (Doughty 2000, 8f.). The smell from the trees is strong and evocative in ways that can prove alienating for some, but at other times and places can serve as a distinguishing feature of home. As one Australian I conversed with while browsing for books on the eucalyptus put it: "In summer or heavy rain, their smell assaults you—in a good way! I know I am home!" It was indeed this olfactory aspect of the eucalyptus that surfaced as a key factor in their early transfers to South Africa, according to the archival testimonies I encountered in King William's Town on the Eastern Cape.

Before I turn to the particular significance that these traits would gain in the South African colonial atmosphere, there is a need to take a brief look at general European sensibilities toward the atmosphere during the latter half of the nineteenth century and beginning of the twentieth.

Domesticating the Atmosphere of Colonial Natures

European settlers in South Africa did not have the privilege of an "unquestionably given and anxiety-free" (Sloterdijk 2009, 47) relation to the air they were breathing. On the contrary, the medical paradigm that dominated the nineteenth century marked colonial tropical atmospheres as highly problematic and a source of anxious attention. This trait, I will argue, dominated certain aspects of settler life in ways that have thus far escaped most academic approaches to colonialism.

The fact that the colonies were often referred to as the "white man's grave" and the "diseased continent" (cf. Comaroff 1993; Curtin 1961, 1989) was due to the tropical atmosphere. Galenic medicine dominated Victorian sensibilities regarding illness and health. Based on humoral theory, the body was viewed as consisting of four fluids, which, if imbalanced, would cause disease: black bile, yellow or red bile, blood, and phlegm. As already mentioned, within such a framework, disease was conceptualized as an atmospheric disturbance, and illness was associated with hot, humid air—a result of breathing miasma, the fumes of rotting, putrefying organic material. Furthermore, since Victorian sensibilities took all strong "smell [to be] disease" (Schoenwald 1973, 681) and pleasant odors to be cures (Miller 1997, 66), the settlers entered a world where they not only distrusted the atmosphere they were breathing but also experienced the quality of the air at times as a matter of life and death. In this world, the nose became an important guide through a bewildering terrain where health and disease were problematic aerial aspects of the tropical landscape (Flikke 2016a).

This foundational understanding of disease challenged the very possibility of creating a home in an alien tropical environment. The preoccupations with miasma inserted a public health space of possible contagion between the settlers and the colonial landscape, making comfortable settlement difficult. This was an aerial space and constituted an atmospheric "casing" where breathing became destiny, since miasma caused an imbalance in the humors that would result in disease and possibly death. Hence, breathing became "too important to continue doing in the open" (Sloterdijk 2014, 961). The colonialists needed to know how and where to move in the tropical landscape in order to avoid breathing miasma and to stay healthy.[3] Maintenance of health was thereby concerned with movement through the landscape and the drawing of maps, as well as a recording of meteorological and topographical observations that Fabian (2000) noted bordered on the obsessive. This effectively turned cartography into an aspect of medicine (cf. Flikke 2003; Osborne 1996, 80). Though these maps laid out safe trails on the ground, attention was directed at currents in the air, carefully plotting paths and establishing settlements in places that would as much as possible avoid hot humid air and the "fetid smell" (Comaroff and Comaroff 1991, 123) of the African bodies.

Though a part of the "dark continent," South Africa was blessed with a Mediterranean climate in most coastal regions, in addition to healthy, dry inland air. As such, South Africa was promoted in England as a health resort during the latter half of the nineteenth century (e.g., Fuller 1892, Scholtz 1879; Marshall-Hall 1908). A noticeable boom in these writings started after the Suez Canal opened in November 1869 and the flow of travelers passing through

South Africa decreased significantly. Though the relations between climate, trees, agriculture, and health were contested in South Africa, as elsewhere (e.g., Bonyhady 2000), Kenneth Thompson (1978, 520) has argued that medical and lay views on public health in America connected breaking new land directly to the release of miasma from the tilled earth, as well. These attitudes remained throughout most of the nineteenth century (Thompson 1978, 517). Viewed within this paradigm, the increased domestication and cultivation of the South African colony directly influenced the otherwise healthy inland atmosphere in ways that needed to be countered. This was clearly addressed in a lecture on "South Africa as a health resort" held in the Royal Colonial Institute in London on November 13, 1888. The speaker, Dr. Symes Thompson, directly linked the danger of breaking new land in the colonies to atmospheric changes that needed to be redressed. Due to miasma being released from the earth, he suggested that settlers plant "a belt of Eucalyptus . . . between the house and the irrigated fields [to act] as an effective screen" (Thompson 1889, 26). There were, in other words, strong sentiments that directly linked colonial agricultural efforts to domesticate colonial lands with destructive atmospheric changes that needed to be rectified.

I have elsewhere (Flikke 2016a) argued that, during a health scare of 1876–77, the town council as well as the citizens of King William's Town addressed the epidemic diseases that ravaged the area through olfactive metaphors that gradually created a public desire to import eucalyptus from the antipodes, since it "absorbed the miasma of towns."[4] The motivating belief behind tree-planting was that eucalyptus, and in particular the Blue Gum (*Eucalyptus globulus*), deodorized and hence sanitized the atmosphere in a town that was described as follows in an editorial in the Cape Mercury: "This afternoon, New Town does not smell, it stinks."[5]

Though the settler towns in the antipodes stunk as well (Wood 2005), there is one big difference between the settlers in Victorian Australia and those in South Africa. The settler communities in King William's Town and Africa in general were largely in a pressed minority situation from the 1870s onward. Furthermore, they were increasingly dependent on a political economy that demanded close contact with, and reliance on, the African majority. King William's Town is special in this regard. It was the site of South Africa's first, and at the time only, native hospital and was at the time treating many Africans suffering from typhoid fever.[6] In addition, it was a frontier town experiencing regular rebellions and armed conflicts. The local prison was therefore filled with native "rebels," and the sanitary condition of the prison surfaced regularly in the papers as a source of "very offensive odours in the rooms, for which no disinfectant had been provided."[7] This gained a new significance as the mineral revolution accelerated. Several studies suggest that times of epidemics increased racial tensions in South Africa. Swanson (1977), for example, traced the first segregationist laws back to public health legislation. Similar traits dominated much of the reactions to the Spanish Flu pandemic of October 1918 (Philips 1984), though the dependence on African labor and the costs of sanitary reforms made most of these social programs short-lived.

Even though King William's Town was small, had a dry and "health-bringing climate" free from malaria, and had a climate comparable to that of Australia, its smell was seen as a problematic aspect of African presence in public spaces. The fact that the town was strategically placed in a conflict zone between European settlers and an African populace ravaged by smallpox epidemics and the consequences of the great Xhosa cattle killing (1856–57) (Peires 1989) increased the potential conflicts associated with the aerial presence of harmful "otherness," be it African nature, people, refuse, or the "deranged," which was another concern that surfaced in the public health debates. Initially, nature and manure, as an aspect of the agricultural economy, generated the most miasmatic and medical concerns and received the bulk of attention in town hall meetings and newspaper editorials. However, my archival investigations reveal an ever-increasing focus on human refuse and the human body, in and of itself, as sources of dangerous odors. The African patients, prisoners, and "mentally deranged" were targeted as sources of epidemic contagion that floated in the air, invisible to the eye but detectable by the nose. It was no longer enough to control the effect of landscapes on the atmosphere. Bodies—and in particular breathing African bodies emitting disease-bringing stenches into the atmosphere—also need to be deodorized, and eucalypts were one mid-Victorian means to address this concern. The King William's Town's Borough Council meeting July 3, 1877, for instance, granted one Mr. Honey the right to plant trees on his property and advised him to plant blue gums, since they "absorbed the miasma of towns." At the same time, the city engineer advised postponing planting on Alexandra Street until it was properly formed.[8]

I focus on how these concerns, which constituted what the medical historians Laidler and Gelfand coined a "sanitation hysteria" (1971, 362), came to call for olfactory air-conditioning. Colonizers worked on the atmosphere to construct a casing for themselves not through physical structures but through deliberately separating healthy and diseased spaces by distributing aromatic traces that enabled them to navigate the treacherous African landscape.

"Air Conditioning" the Colonial Domus

As such, I claim that the health scare in King William's Town illustrates a particular case of what Peter Sloterdijk has coined "air conditioning," understood here as "disconnecting a defined volume of space from the surrounding air" (2009, 20). It is not my intention here to compare the European colonial interaction with the atmosphere with the terror of poison gas exterminations of the twentieth century. Yet Sloterdijk brings up one important concern I will push, namely, the acute uncertainty that can potentially be inserted between humans and the atmosphere as the primary source of life. Wars are won or lost based on the capacity to conquer the respiratory potential of hostile people or climates. For the Victorians, breathing the tropical air was a question of destiny, and air conditioning—understood here as attempts to control heat, humidity, and stenches—was potentially a question of life and death (cf. Sloterdijk 2014, 961ff.).

Following Sloterdijk, I have suggested that in studies of domestication and the colonial acclimatization, we have much to gain analytically from more closely interrogating the human atmosphere dependency and considering the existential uncertainty involved in moving about in a landscape in which you do not trust the atmosphere's capability to sustain life. The historical material I discuss in this chapter clearly indicates that the settlers did not have an unquestionable, trusting, and anxiety-free attitude toward colonial "being-in-the-air," as Sloterdijk (2009, 48) paraphrased Heidegger. I have argued that the health scare in King William's Town reveals a similar sense of existential distrust of the air the settlers were breathing, because it was infused with forms of "otherness" associated with disease and revealed through smell. As an existential reality, the regular epidemics of typhus, cholera, flu, smallpox, and TB turned the air into potential poison and made it unsuitable as a home. In such an alien and hostile environment, the planting of eucalyptus domesticated the air of the colonial contact zone enough for it to appear safe, orderly, and sanitized, thus providing an aerial domus in the tropics.

What I have outlined constitutes a colonial contact zone where the need to acclimatize both species and people to new landscapes was viewed as a continuation of a larger colonial domestication project, which included the air and atmosphere. Through planting eucalyptus, the settlers worked on the air, adapted themselves to the colonial atmosphere, and created new ecological and social fields where sensory engagements with the natural and social landscape rooted them in the colonial peripheries, literally grafting themselves into the landscape as they planted eucalypts in King William's Town to secure a future there. One unexpected convergence centered on olfaction and the way it changed the colonial landscapes as well as the health-seeking behavior of the settlers.

I have so far used "air" and "atmosphere" as synonyms and will now argue that the transplants of eucalypts I have accounted for were conceived of as literally changing the colonial atmosphere, understood as ambience. In this sense, the domestication of colonial species and the refashioning of landscapes through the planting of deodorizing eucalyptus influenced the air in ways that created a sense of recognition and belonging, which helped the settlers find a home in their new country.

Atmospheres as Concept—Smell as Experience

Despite the fact that I will deal with atmosphere in its original sense as the mass of air enveloping the earth,[9] there is a need to consider its metaphorical extension as ambience. In an essay on atmosphere, Gernot Böhme (1993) wants to situate atmosphere as a fundamental concept in a theory of aesthetics. Böhme argued that a taken-for-granted subject-object dichotomy positioned atmosphere as a quality of the subject in ways that prevented the development of an aesthetic theory of atmosphere; acting subjects created ambience and it could hence not be an aspect of objects. Böhme moved away from this perspective to open a space for "things and their ecstasies" (1993, 120), arguing that there is a need to relocate atmospheres ontologically, as occupying spaces between subjects and objects. This can only be done if we stop treating things and objects through their closures and rather trace their connections and spatial relations. As such, atmospheres are spaces "'tinctured' through the presence of things, of persons or environmental constellations" (Böhme 1993, 121). In this perspective, atmospheres are emotional spaces that emerge from the relations between persons, things, and their constellations. Atmospheres, then, are in Böhme's vocabulary not things but thinglike; neither are they subjects but subjectlike. They are experienced as a bodily state of being localized in subjects but must be accounted for in relation to a larger field of material relations.

I will suggest that the atmosphere we breathe and move through has some of the same challenges regarding its ontological positioning. In most anthropological studies, the atmosphere is mostly treated as an empty space through which autonomous subjects move and sensorily experience the external world. Along the lines of Böhme, I suggest that the materiality of atmospheres, such as the air we breathe and the space we move through, has the same transgressive ontological qualities as ambience—it is external yet inseparable from the subject. The moment I stop inhaling and exhaling through my lungs and skin, I am no longer a subject but an object—a corpse. Ontologically speaking, atmosphere as air is hence inseparable from subjects and needs to be considered as part of processes of subjectification. The atmosphere and air we breathe is therefore not only a part of our external material surroundings but is integral to lived life and as such part of self.

This has some interesting implications for my argument regarding the olfactive traces of eucalypts that started to spread in King William's Town during the mid-1870s. As processes of subjectification, the air we breathe and the aromas enveloping us are part of self and can hence not be separated from the breathing subject as such. When the citizens of King William's Town were breathing in air deodorized by eucalypts, they were breathing health, well-being, and prosperity. I have elsewhere (Flikke 2016a) argued that neuro- and evolutionary biologists have pointed out that to smell is literally to be touched and thus to bring the distant source of odors in immediate contact with the sensing subject when molecules from the subject or object smelled actually touch and blend with our olfactory bulbs. Furthermore, unusual smells trigger emotions and reactions that bypass conscious processing. It is the emotional fervor attached to smells that I argue locates olfaction as an engine in social processes intimately related to the colonial conquest of southern Africa as an aspect of the domestication of the Dark Continent. The smell of eucalyptus, which could be alienating for settlers in the antipodes who needed to clear pastures to secure their future, evoked different emotional responses when encountered in King William's Town, which was regularly plagued by epidemics throughout most of the nineteenth century. What I suggest is that recent scientific studies on olfaction—which underscore that strong, unusual smells in situations of distress are likely to be accompanied by strong emotions—can help shed light on the continued efforts to domesticate eucalyptus in the South African environments even as Australian settlers preferred to import and plant oak and pine trees. Furthermore, it clearly shows that it was not solely plants and animals being domesticated and rooted in new environments; people needed to be acclimatized to new atmospheres, as well. The medical paradigm singled out eucalyptus as an attractive means to secure the settlers' existence in their new territories. Domestication, I argue, is indeed a helpful concept for exploring these connections.

Conclusion: Domesticating the Unknown

In this chapter, I have argued that the colonial domestication of air and atmosphere as it unfolded in Southern Africa is a gateway to new insights into the colonial contact zone. By emphasizing the material aspects of air and connecting it to the Victorian ethos, I have argued that the domestication of air was a necessary aspect of successful acclimatization for the European settlers in the tropics. As such, I have suggested that the gradual domestication of air is an understudied aspect of the colonial conquest with some interesting and far-reaching ramifications. I focused first on the rising popularity of the eucalypts among settlers in the African colonies and argued that the atmospheric aspects of the trees could explain their rapid expansion throughout the world. At the same time, atmosphere as ambience was used to account for the contemporary unpopularity and highly contested status of eucalypts on their home turf in the antipodes.

I emphasized how the archive material underscored the olfactive presence of eucalyptus and argued that the scent of the eucalypt could be traced across the contemporary understanding of public health and contagion. This occurred at the time medicine was revolutionized by the discovery of the biomedical disease vector. The changing understanding of contagion spatially repositioned the critical public health space, fueling a growing fervor to sanitize interracial spaces through olfactive air-conditioning. I concluded this section by arguing that the Victorian epidemiological connection between smell and health suggests that the introduction and production of eucalyptus should be understood as a way the settlers domesticated an uncertain colonial atmosphere in ways that improved their chances for a good life in an otherwise hostile environment, ripe with poverty, suffering, and disease. This existential focus can explain the different sensibilities the settler communities—which, after all, were recruited from much the same English milieus—displayed to eucalypts in South Africa and in the antipodes.

In an effort to strengthen this part of my argument, I outlined recent biological research on olfaction. This research shows that scents carried through the air ontologically change the atmosphere. To smell is physiologically a process that materially merges subject and object. The threats of the distant new and unknown are brought close. The particular neurological processing of olfaction promises to add another layer to our understanding of the colonial contact zone and human efforts to come to terms with otherness in general. My argument suggests that the domestication of foreign, unpredictable, and threatening colonial atmospheres through transplants of trees is also a way of domesticating the negative affects—atmosphere—created by strange smells as existential threats.

NOTES

1. In 1876, at the outset of the "sanitation hysteria" (as the King William's Town episode has been labeled), the population was 3,242, and 489 were reported dead. This was a death rate of 17.41 per 1,000, excluding Africans (*Cape Mercury*, April 25, 1877). This intensified in the following months, and ninety people were buried in April 1878 (editorial, *Cape Mercury*, May 15, 1878).

2. *Cape Mercury*, Monday, April 16, 1877, 3.

3. This affected the colonial conquest in a number of related ways that I cannot fully discuss here, such as locating trade centers at high altitudes away from the humid and "malarial" harbors (Curtin 1985), as well as the development of colonial architecture, including the invention of the verandah, elevated from the hurly-burly of the unsanitary streets, and well ventilated by the winds (cf. Myers 2003).

4. *Cape Mercury*, Wednesday, July 4, 1877, 2.

5. Editorial from the *Cape Mercury*, Wednesday, May 15, 1878.

6. See, for instance, the *Cape Mercury*, February 27, 1878, letter to the editor signed by SANITAS.

7. The doctor's report, published in the *Cape Mercury*, April 3, 1878.

8. *Cape Mercury*, Wednesday, July 4, 1877.

9. From the Greek: *athmós*, "vapor," and *sphaîra*, "ball" or "globe."

REFERENCES

Armstrong, David. 1993. "Public Health Spaces and the Fabrication of Identity." *Sociology* 27, no. 3: 393–410.

Arnold, David, ed. 1996. *Warm Climates and Western Medicine: The Emergence of Tropical Medicine, 1500–1900*. Ed. W. F. Bynum and R. Porter, the Welcome Institute Series in the History of Medicine. Amsterdam and Atlanta, GA: Rodopi.

Beinart, William, and Karen Middleton. 2004. "Plant Transfers in Historical Perspective: A Review Article." *Environment and History* 10: 3–29.

Benjamin, Walter. 1999. *The Arcades Project*. Trans. H. Eiland and K. McLaughlin. Cambridge, MA: Belknap.

Bennett, Brett M. 2010. "The El Dorado of Forestry: The Eucalyptus in India, South Africa, and Thailand, 1850–2000." *International Review of Social History* 55, no. S18: 27–50.

Bennett, Brett M. 2011. "A Global History of Australian Trees." *Journal of the History of Biology* 44: 125–45.

Bennett, Brett M. 2014. "Model Invasions and the Development of National Concerns over Invasive Introduced Trees: Insights from South African History." *Biological Invasions* 16: 499–512.

Böhme, Gernot. 1993. "Atmosphere as the Fundamental Concept of a New Aesthetics." *Thesis Eleven* 36: 113–26.
Bonyhady, Tim. 2000. *The Colonial Earth*. Melbourne: Melbourne University Press.
Brockway, Lucile H. 2002. *Science and Colonial Expansion: The Role of the British Royal Botanic Gardens*. 2nd ed. New Haven: Yale University Press.
Burgess, G. H. O. 1968. *The Eccentric Ark: The Curious World of Frank Buckland*. New York: Horizon.
Butchart, Alexander. 1998. *The Anatomy of Power: European Construction of the African Body*. New York: Zed Books.
Choy, Timothy. 2011. *Ecologies of Comparison: An Ethnography of Endangerment in Hong Kong*. Durham: Duke University Press.
Choy, Timothy. 2012. "Air's Substantiations." In *Lively Capital: Biotechnologies, Ethics, and Governance in Global Markets*, ed. K. S. Rajan. Durham: Duke University Press.
Choy, Timothy, and Jerry Zee. 2015. "Condition—Suspension." *Cultural Anthropology* 30, no. 2: 210–23.
Comaroff, Jean. 1993. "The Diseased Heart of Africa: Medicine, Colonialism, and the Black Body." In *Knowledge, Power and Practice: The Anthropology of Medicine and Everyday Life*, ed. S. Lindenbaum and M. Lock. Berkeley: University of California Press.
Comaroff, Jean, and John L. Comaroff. 1991. *Of Revelation and Revolution: Christianity, Colonialism, and Consciousness in South Africa*. Vol. 1. Chicago: University of Chicago Press.
Comaroff, Jean, and John L. Comaroff. 1997. *Of Revelation and Revolution: The Dialectics of Modernity on a South African Frontier*. Vol. 2. Chicago: University of Chicago Press.
Comaroff, Jean, and John L. Comaroff. 2001. "Naturing the Nation: Aliens, Apocalypse and the Postcolonial State." *Journal of Southern African Studies* 27, no. 3: 627–51.
Corton, Christine L. 2015. *London Fog: The Biography*. Cambridge, MA: Harvard University Press.
Crosby, Alfred. 1993. *Ecological Imperialism: The Biological Expansion of Europe, 900–1900*. Canto Classics. Cambridge: Cambridge University Press.
Crosby, Alfred. 2004. *Ecological Imperialism: The Biological Expansion of Europe, 900–1900*. Ed. D. Worster and J. R. McNeill. With a new preface, 2nd ed., Studies in Environment and History. Cambridge: Cambridge University Press. Original edition, 1986.
Curtin, Philip D. 1961. "The White Man's Grave: Image and Reality, 1780–1850." *Journal of British Studies* 1, no. 1: 94–110.
Curtin, Philip D. 1985. "Medical Knowledge and Urban Planning in Tropical Africa." *American Historical Review* 90: 594–613.
Curtin, Philip D. 1989. *Death by Migration: Europe's Encounter with the Tropical World in the Nineteenth Century*. Cambridge: Cambridge University Press.
Diamond, Jared. 1997. *Guns, Germs, and Steel: The Fates of Human Societies*. New York: W. W. Norton.

Donner, Simon E. 2007. "Domain of the Gods: An Editorial Essay." *Climatic Change* 85: 231–36.
Dunlap, Thomas R. 1997. "Remaking the Land: The Acclimatization Movement and Anglo Ideas of Nature." *Journal of World History* 8: 303–19.
Fabian, Johannes. 2000. *Out of Our Minds: Reason and Madness in the Exploration of Central Africa*. Berkeley: University of California Press.
Feierman, Steven, and John M. Janzen, eds. 1992. *The Social Basis of Health and Healing in Africa*. Berkeley: University of California Press.
Flikke, Rune. 2003. "Public Health and the Development of Racial Segregation in South Africa." *Bulletin of the Royal Institute for Inter-Faith Studies* 5, no. 1: 5–23.
Flikke, Rune. 2016a. "South African Eucalypts: Health, Trees, and Atmospheres in the Colonial Contact Zone." *Geoforum* 76: 20–27.
Flikke, Rune. 2016b. "Enwinding Social Theory: Wind and Weather in Zulu Zionist Sensorial Experience." *Social Analysis* 60, no. 3: 95–111.
Franklin, Adrian. 2006. "Burning Cities: A Posthumanist Account of Australians and Eucalypts." *Society and Space* 24: 555–76.
Fuller, Arthur. 1892. *South Africa as a Health Resort: With a Special Reference to the Effects of the Climate on Consumptive Invalids, and Full Particulars of the Various Localities Most Suitable for their Treatment, and also of the Best Means of Reaching the Places Indicated*. 1898, 6th ed. London: Union Steam Ship Company.
Hay, Ashley. 2002. *Gum: The Story of Eucalypts and Their Champions*. Potts Point, NSW: Duffy and Snellgrove.
Hirschfeld, Christian C. L. 2001. *Theory of Garden Art*. Trans. L. B. Parshall. Ed. J. D. Hunt, Penn Studies in Landscape Architecture: University of Pennsylvania Press. Original edition, Theorie der Gartenkunst, 1779–1785.
Hulme, Mike. 2014. "Better Weather? The Cultivation of the Sky; Essay for 'Openings and Retrospectives: Life above Earth.'" *Cultural Anthropology* 30, no. 2: 236–44.
Ingold, Tim. 2010. "Footprints through the Weather-World: Walking, Breathing, Knowing." *Journal of the Royal Anthropological Institute* 16 (Supplement s1): s121–s139.
Ingold, Tim. 2011. *Being Alive: Essays on Movement, Knowledge and Description*. London and New York: Routledge.
Jenkins, Wills. 2005. "Assessing Metaphors of Agency: Intervention, Perfection, and Care as Models of Environmental Practice." *Environmental Ethics* 27, no. 2: 135–54.
Kirksey, S. Eben, and Stefan Helmreich. 2010. "The Emergence of Multispecies Ethnography." *Cultural Anthropology* 25, no. 4: 545–76.
Koslofsky, Craig. 2011. *Evening's Empire: A History of the Night in Early Modern Europe*. Cambridge: Cambridge University Press.
Laidler, Percy Ward, and Michael Gelfand. 1971. *South Africa: Its Medical History, 1652–1898*. Cape Town: C. Struik.
Lever, Christopher. 1992. *They Dined on Eland: The Story of the Acclimatisation Societies*. London: Quiller.
Lien, Marianne Elisabeth. 2007. "Weeding Tasmanian Bush: Biomigration and Land-

scape Imagery." In *Holding Worlds Together: Ethnographies of Knowing and Belonging*, ed. Marianne. E. Lien and Marit Melhuus. Oxford: Berghahn.
Lien, Marianne Elisabeth, and Aidan Davison. 2010. "Roots, Rupture and Remembrance." *Journal of Material Culture* 15, no. 2: 1–21.
Marshall-Hall, A. S. 1908. *Cape Colony: The Land of Sunshine and Health.* Cape Town: Cape Government Railway Department.
Miller, William Ian. 1997. *The Anatomy of Disgust.* Cambridge, MA: Harvard University Press.
Myers, Garth Andrew. 2003. *Verandahs of Power: Colonialism and Space in Urban Africa.* Ed. J. R. Short, in the Space, Place, and Society series. Syracuse, NY: Syracuse University Press.
Osborne, Michael A. 1996. "Resurrecting Hippocrates: Hygienic Sciences and the French Scientific Expeditions to Egypt, Morea and Algeria." In *Warm Climates and Western Medicine: The Emergence of Tropical Medicine, 1500–1900*, ed. D. Arnold. Amsterdam and Atlanta, GA: Rodopi.
Osborne, Michael A. 2001. "Acclimatizing the World: A History of the Paradigmatic Colonial Science." *Osiris* 15: 135–51.
Peires, Jeffrey B. 1989. *The Dead Will Arise: Nongqawuse and the Great Xhosa Cattle-Killing Movement of 1856–7.* Johannesburg: Ravan.
Pelling, Margaret. 1978. *Cholera, Fever, and English Medicine, 1825–1865.* Oxford: Oxford University Press.
Phillips, Howard. 1984. "Black October: The Impact of the Spanish Influenza Epidemic of 1918 on South Africa." PhD thesis, University of Cape Town.
Pyne, Stephen J. 1998. *Burning Bush: A Fire History of Australia.* With a foreword by William Cronon and a new preface by the author. Seattle: University of Washington Press.
Sargent, Frederick, II. 1982. *The Hippocratic Heritage: A History of Ideas about Weather and Human Health.* New York: Pergamon.
Schoenwald, Richard L. 1973. "Training Urban Man: A Hypothesis about the Sanitary Movement." In *The Victorian City: Images and Realities*, ed. H. J. Dyos and M. Wolff. London: Routledge and Kegan Paul.
Scholtz, William C. 1897. *The South African Climate.* London: Cassell.
Sim, T. R. 1905. *Tree Planting in Natal.* Vol. 7, *Bulletin Natal Department of Agriculture.* Pietermaritzburg: P. Davis and Sons.
Sloterdijk, Peter. 2009. *Terror from the Air.* Trans. A. Patton and S. Corcoran, in the Foreign Agents series. Los Angeles: Semiotext(e). Original edition, 2002, as *Luftbeben.*
Sloterdijk, Peter. 2014. *Spheres II: Globes, Macrospherology.* Trans. W. Hoban. English translation, in the Foreign Agents series. Los Angeles: Semiotext(e). Original edition, Sphären I. Blasen.
Swanson, Maynard W. 1977. "The Sanitation Syndrome: Bubonic Plague and Urban Native Policy in the Cape Colony, 1900–1909." *Journal of African History* 18, no. 3: 387–410.

Thompson, Kenneth. 1978. "Trees as a Theme in Medical Geography and Public Health." *Bulletin of the New York Academy of Medicine* 54, no. 3: 517–31.
Thompson, Symes. 1889. "South Africa as a Health Resort." In *Report of Proceedings*, ed. R. C. Institute. London.
Wood, Pamela. 2005. *Dirt: Filth and Decay in a New World Arcadia*. Auckland: Auckland University.
Zacharin, Robert Fyfe. 1978. *Emigrant Eucalypts: Gum Trees as Exotics*. Melbourne: Melbourne University Press.
Zee, Jerry. 2015. "Breathing in the City: Beijing and the Architecture of Air." *Scapegoat: Architecture, Landscape; Political Economy* (Winter/Spring): 46–56.

9

HOW THE SALMON FOUND ITS WAY HOME

Science, State Ownership, and the Domestication of Wild Fish · *Gro B. Ween and Heather Anne Swanson*

Classic definitions of domestication have typically stressed the importance of physical confinement for the remaking of animal kinds. Enclosure within fences and pens is often depicted as a change that not only allows for more reproductive control over animals but also ushers in new forms of ownership and property claims. In a general understanding of domestication, isolation from the wild and the introduction of mechanisms that enable ownership are essential. This chapter acknowledges such strong links between domestication, enclosure, and ownership, but—in line with the thrust of this volume—it argues that we must broaden our definitions of domestication to look at forms of human-animal relations that do not manifest as literal confinement or strict reproductive control. Through attention to migratory Atlantic salmon, this chapter explores how animals—even those we consider to be "wild"—can become domesticated and owned, not through isolation or physical enclosure but through incorporation into "domestic" imaginaries of nation-states. How, we ask, do wild salmon become a "resource" enclosed within the "domus" of the nation-state? How do such broader forms of domestication also remake salmon-human relations?

With the rise of pen-based aquaculture, many Atlantic salmon have been thoroughly domesticated in the narrow sense of the term. They have been held captive and made subject to intensive breeding and pellet diets (Lien 2015). Despite the fish-farming boom, sizable numbers of Atlantic salmon, *Salmo salar*, continue to live "wild" lives, migrating in the North Atlantic Ocean and spawning on their own in European rivers. Although classified as "wild," these salmon do not escape domestication and its logics. Even in the most remote areas, they have been brought into regimes of nation-state ownership that have themselves emerged out of classic domestication narratives.

Today, governments on the edges of both the Atlantic and the Pacific make strong ownership claims to salmon spawning in national rivers as well as to salmon swimming in the commons of the ocean (see, for example, Grey 1998; Symes and Phillipson 2001). As we will show, fish biology has made it possible for nation-states to claim salmon as "Norwegian" or "American" or "Japanese" even when they are swimming on the high seas beyond the limit of their territorial waters.[1] This notion of salmon as being legally bound to a political entity regardless of their location is new in the history of human-salmon relations. It gradually came into being over the past century through the articulation of existing ideas of state resource control alongside new scientific research on salmon homing and migration. Before they could be "homed in" to the nation-state, salmon needed to be understood as having a "home" that tied them to particular lands in the first place. This chapter traces the development of this resource ontology (Richardson and Weszkalnys 2014)—one that roots salmon to particular places—as it emerges out of new scientific knowledges about salmon homing and migration. These knowledges reconfigured ideas about salmon belonging and ownership by making it possible to identify the migration routes and natal origins of salmon when they are swimming in the ocean.

While inquiries into salmon migration and homing emerged from basic scientific curiosity, they later became central to the imposition of a worldview—one that sought to nationalize salmon in historically novel ways (Pálsson 1998). In the following sections, we investigate a series of key texts in fish biology that show how this nationalized salmon was produced. These scientific endeavors sought to better understand fish migration, initially through simple technologies, such as threads around fish tail fins, and more recently through complicated tracking devices, such satellite tags in salmon bellies. Through such tracking and fish identification practices, scientists asked a range of linked questions: Was it the same salmon that appeared in the same river year after year? Where did the salmon go when they were not there? Why do salmon migrate? And how, exactly, do they find their way home? The answers to such queries generated new senses of salmon homing, theories of salmon homing, and maps of the routes that salmon travel. These are the very knowledges, we argue, that facilitated the incorporation of wild salmon into the nation-state "domus."

Homing and Returns

"Home" is undoubtedly one of the most important concepts in popular stories about salmon. The salmon is a creature who always returns home, often against great obstacles. After years of ocean exploration, it struggles against the currents and leaps up waterfalls, a faithful offspring who remembers its origins and remains dedicated to its homeland. While a salmon typically spends more time in saltwater than in fresh, a salmon's "home" is unquestionably depicted as the river where it is born and dies, the place to which it has enduring ties. The scientific language of salmon homing conjures similar sensibilities about "home" and the importance of ties to one's birthplace. "Stream fidelity" is a technical term that describes the rates at which populations return to their rivers, while "straying" is a term that describes the behavior of those individuals who do not.[2] Synonyms for homing behavior include "philopatry," a term coined in the 1940s out of "philo" or love and the Greek *patra*, for fatherland. These descriptions are shot through with European tropes of "home" and nation-state belonging.[3]

Although such stories of salmon "homing" are now iconic of the fish, this was not always the case. For example, the Sámi Veide (coastal Sámi) of northern Norway have long focused on the fish's *return* rather its sense of home. As Ingold (2000) has pointed out, for the Sámi Veide, trust was likely the main point of significance in human-animal relations: "The trust that situates humans and animals in a relational weave, entangled and mutually depended upon each other" (70). People could observe that fish ready to spawn returned to rivers when the ice broke, new young ones appeared, and both the young and the old traveled down river in the spring. The return of the fish was ensured by the reciprocal and respectful behavior of salmon, the river, and other relevant cospecies. Through care and correct behavior, Sámi people could trust that fish would reappear the following year (Ween 2012a).

As early as the sixteenth century, however, some more southerly Europeans were beginning to imagine rivers as a salmon "home." Two clergymen, Hector Boece, a Scot, and Peder Claussøn Friis, a Norwegian, independently theorized about the travels of the Atlantic salmon and depicted the river as their birthplace (Nordeng and Bratland 2006). Friis, the Norwegian, hypothesized that Atlantic salmon, in their seasonal migrations, were guided by a leader, a so-called pilot fish, that had God-given supernatural abilities (Nordeng and Bratland 2006). Friis proposed that the salmon were guided by the kingfish (a pelagic fish now known as *Lampris guttatus*), who was locally referred to as the "salmon steerer." What was "most to wonder at," according to Friis, "is that salmon still find their way back to the home river, and the actual place where they were born and bred, to reproduce their stock and species" (Nordeng and Bratland, 2006, 489). In stressing the river as the home, where fish were "born and bred," Friis's language was already setting the stage for the notions of fish nationality that emerged more fully in later centuries.

The eighteenth century saw a growing scientific interest in animal migrations, including those of birds and fish. Many early scientific works, however, did not see animal disappearances and reappearances as journeys. Aristotle believed that birds transformed into different species to overwinter, while Linnaeus believed that swallows disappeared into swamps in the autumn and reappeared again in the spring (Berthold 1993, 11). But as early as 1702, migration began to emerge as a scientific concept. In that year, von Pernau, an Austrian count with a passion for ornithology, described the migration of birds. His studies of bird migration were based on observation; he noted that the restlessness of caged birds coincided with the timing of wild birds' annual flights. Von Pernau attributed this restlessness to an instinct, "a hidden drive at the right time" (Berthold, 1993, 12). Other scientists' research in the eighteenth century included experiments in attaching rings to birds' legs to identify if those who returned were the same as those who left (Dingle 1996). Similar experimental efforts to understand salmon journeys occurred around the same time. As was the case with the bird leg rings, the technology employed was simple. Juvenile salmon were marked with thread tied around their tails. If the threaded salmon reappeared in the same river in the coming years, scientists possessed powerful evidence that the salmon had returned to its natal river (Quinn 2005).

Making "Homing"

Although an awareness of fish movement gradually emerged as European scientists recognized that thread-marked fish returned, a handful of small-scale experiments could not establish the existence of salmon homing as a species-wide phenomenon or as a distinguishing trait. From the earliest experiments, which involved very small numbers of fish, scientists did not know if all salmon returned to the streams of their birth or if only a few did. In a chapter about the history of fisheries science, Quinn, a noted American fisheries biologist, describes the development of homing experiments from the simple thread studies to large-scale marking projects that aimed to confirm that most fish return to their natal rivers (2005). These later experiments were not possible until the early to mid-twentieth century, when new technologies, such as the hatchery production of Pacific salmon and better tagging technologies, came into play (Hoare 2009). Among the first mass-marking experiments was a 1931 project that clipped the ventral fins of 362,265 sockeye salmon smolts in British Columbia's Lower Fraser River system—a substantial undertaking (Foerster 1954, 1968). The fish were expected to return two years later. In 1933, the project counted 2,856 large marked salmon, along with 409 small unmarked fish and 17 unmarked large ones.

The results presented new unknowns. The scientists were confident that they had an explanation for the small fish: these were fish who had spent only one year at sea and thus had not been part of the marked cohort. But what about the seventeen unmarked large fish? The scientists speculated that their fins might have grown back or that, perhaps, they had escaped the marking. Not entirely happy with the results, the experiment's lead scientist repeated the experiment again in 1936, this time clipping the fins of 497,598 smolts. The new experiment resulted in even larger numbers of unmarked fish, both small and large: 8,980 large marked fish returned, along with 4,226 small, unmarked fish, and 136 large unmarked fish. Regardless of the unexplained presence of large fish with intact fins, however, the study confirmed that the fish could be counted on to return to their origin.

After the British Columbia experiment, fin-clipping became a common method of marking and tracking fish. The method had limitations: fin-clipping substantially increased the mortality of fish at sea (Foerster 1954), and there were continued concerns that fins could grow back (Quinn 2005). Furthermore, because there are only so many combinations of fins one can clip, fin-clipping limited the number of different experiments that could take place in the same area at the same time. If multiple projects used similar marks, it was hard to tell which fish had been clipped as part of which project (2005, 86).

By helping to confirm widespread salmon homing behavior, fin-clips increased the perception that salmon were linked to specific rivers and that the rivers were their "homes." But those identities didn't yet travel with the fish; when salmon swam in the ocean they remained anonymous and stateless. Fin-clipping was not an effective tool for identifying salmon in the ocean and thus limited scientists' abilities to study fish migratory patterns and routes. Significant numbers of scientists were fin-clipping salmon in many rivers in different countries. Thus, on the rare occasion that a fisherman captured at fin-clipped fish sea, he would have no idea who to contact, and even if he did, no one could tell with confidence from which marking project the fish had originated.

The use of fish tags, which began in the 1940s, dramatically improved fish identification in the ocean. These early tags were small metal plaques that could be tied to the dorsal fin with a fine metal thread. The tag had a number on one side and, on the other, an address of the government agency to which the plaque should be returned (Quinn 2005). No matter where the fish was found, the tag unambiguously connected the fish to its homeland. It also transformed the sea into a new salmon research site. On June 17, 1940, tagged Atlantic salmon were caught west of Newfoundland. The tag told researchers that the fish had originated in the Margaree River in Breton, France, in 1938. On September 21, 1940, the same fish were recovered in their home river, after having traveled more than nine hundred kilometers (Quinn 2005, 86). Soon, scientists around the world began similar inscriptions of salmon on a large scale. As early as the 1950s, biologists established that fish tagged in rivers along the U.S. East Coast were caught in the great Greenland fisheries (Report of ICES/ICNAF 1969; Jensen 1968, in Stasko 1971).

Still, the tagging method had its uncertainties. Fastening metal plaques around dorsal fins with a wire was time-consuming. Such tagging systems were also vulnerable, as the tags could easily fall off. The coded wire micro-tag, which is carried internally, addressed this issue and vastly improved migration studies (see, for example, Jefferts et al. 1963). Developed in the 1960s, the micro-tag is—as its name implies—very small; it consists of a 1mm coil of fine stainless steel wire initially with a binary code, and later a decimal code, etched onto it. Instead of being externally attached, the micro-tag is inserted into the nasal cartilage of a juvenile salmon. The codes were complex enough to give unique identification numbers first to hundreds of thousands of small fish groups, and later to millions of individual fish. These identification codes allowed scientists to map the ocean routes of salmon populations in new ways. Like the earlier plate tags, micro-tags contained information about the river from which a fish originated. But in contrast to earlier tags, micro-tags were much cheaper to manufacture and could be inserted into much smaller and younger fish. While plate tags were so bulky that they posed an intolerable burden for small juvenile fish, the micro-tags did not interfere with their swimming.

The extensive hatchery production of Pacific Salmon (Taylor 1999) also increased the logistical ease of fish-tagging experiments along the U.S. and Canadian west coasts. Hatcheries produced large numbers of juvenile fish and then released them into rivers. Those fish subsequently migrated to the ocean and then back to the rivers of their release. Hatchery rearing of salmon eliminated one of the challenges of fish marking: one did not have to go out and catch juvenile fish to mark them. The combination of the technical ease of micro-tags and large numbers of hatchery fish available to be marked dramatically shifted the scale of tagging experiments from thousands of fish to millions of them. For example, between 1974 and 1977, in the Cowlitz River, a single tributary of Columbia River, one research project tagged 1.2 million smolts (Quinn and Fresh 1984, in Quinn 2005).

Micro-tags, however, are not without limitations. If they are not inserted properly, they can wander within the fish body and either work their way out through the fish's skin or even kill the fish (Quinn 2005). They are also relatively hard to retrieve from sports and commercial fisheries (Jefferts et al. 1963). First, without additional markings, it is impossible to tell which fish contain micro-tags without the use of a metal detector. Scientists thus clip the adipose fins of all micro-tagged fish to indicate which carry tags. With this external mark, fishermen can identify which salmon carry micro-tags, but it remains difficult for a layperson to dig the tiny tags out of fish snouts. To address this problem, scientists often collect entire salmon heads at docks and extract the tags themselves. While these tags are now magnetized to ease their retrieval, scientists are still only able to retrieve a small percentage of the micro-tags they implant. After retrieval, researchers read the codes on micro-tags under a microscope to learn from which hatchery the fish originated and of which experimental group it was part (Quinn 2005, 88). By the 1980s, such tags became so common that hatcheries routinely examined all returning adult salmon for tags (Quinn 2005, 89).

One of the important contributions of micro-tags was the data they produced when they were collected at sea. When fishermen caught a micro-tagged fish in the open ocean, scientists could confirm the exact origin of that fish. This allowed them to map ocean migratory routes of fish populations in much more precise ways. While plate tags had offered the same theoretical possibilities, they were so expensive and such a limited number of fish carried them that retrieving them in the ocean was akin to finding the proverbial needle in a haystack. With millions of micro-tags, the possibility of finding some tagged fish increased exponentially, allowing for substantial data collection at sea for the first time.

This new data allowed for important insights: scientists realized that salmon born in the rivers in one country were often caught in the waters of another. Such data raised a host of questions about fish ownership. Were nations entitled to catch the salmon that swim in the oceans off their coasts, regardless of their origin? Or did fish, who now carried micro-tags bearing their location of origin, belong to the regions where they had been born and the hatcheries that had labored to produce them? These new debates were brought into being by the micro-tags that allowed salmon to keep their nationalized river identities on the high seas.

Other tagging technologies furthered research on salmon movements in the ocean and amplified such debates. Due to their implantation in fish snouts, micro-tags—and the data they contain—can be retrieved from fish only after they have been killed. They thus allow scientists to identify salmon only one time and in only one place—in the context of a lethal harvest. To better understand salmon *routes*—and not merely where in the ocean specific populations of salmon could be found—scientists needed to be able to track living fish as they moved through time and space.

Fish telemetry—the real-time tracking of swimming fish—first became possible in the late 1950s, with the introduction of ultrasonic transmitters that could be externally attached to mature fish. Such devices allowed fish biologists to track tagged fish by following them in a boat with a hydrophone or to detect them when they passed by fixed-site receivers. But early telemetry tags were so large that they could be attached only to adult fish, making it impossible to study the movements of juveniles (Hockersmith and Beeman 2012). Furthermore, tag detection ranges were limited, meaning that receivers had to be close to the fish to detect them. These factors meant that most telemetry projects were conducted in rivers, where it was easiest to keep track of the fish. Scientists did attempt ocean studies (see Hockersmith and Beeman 2012), but these typically were limited to the short periods during which the scientists were able to follow the fish by boat.

Since the 1980s, acoustic telemetry technologies have dramatically improved. Smaller devices cause considerably less discomfort and allow fish to move more freely, while the ranges of tags and detection equipment facilitate more effective tracking. Today, for example, scientists are able, within a limited area, to locate undelivered tags inside fish that are stored in individual sports fishermen's freezers (Niemelä, pers.com). Most important in the context of this chapter is how telemetry has enabled new insights about fish movements at sea. Since the early 2000s, extensive arrays of telemetry receivers have been placed along the seafloor of both the Pacific and the Atlantic continental shelves (see, for example, the Pacific Ocean Shelf Tracking Project[4] and the

SALSEA program[5] in the Atlantic). These arrays, often located near national boundaries (such as that between Washington state and British Columbia), have allowed scientists to more precisely determine how fish from different natal rivers move back and forth across the marine fishing grounds of different countries, providing additional evidence about who might be catching fish from which rivers of origin.[6]

In recent years, scientists have deployed ever more extensive methods of identifying fish and mapping salmon movements through the use of scale samples, catch statistics, video counters, and new marking techniques (NASCO Fisheries Management Focus Area Report 2007; Ween 2012b). New genetic technologies have been especially important, because they can identify a fish's home tributary through analysis of the genetic material from a single scale (Svenning et al. 2015). Genetic analysis serves identification functions similar to those of micro-tagging, but with added benefits. First, it allows the identification of the home rivers of all fish without active tagging. This is key for identifying the origins of wild fish caught in ocean fisheries, as wild juveniles are difficult to collect and thus rarely carry micro-tags. Second, in contrast to wire micro-tags, genetic analysis doesn't require killing fish; fish can be released alive after a scale has been removed. Through these advantages, this technology has dramatically expanded the ability to identify the river origins of fish in the ocean.

Smell, Vision, Magnetic Fields, and Celestial Bodies

At the same time that they have more precisely mapped salmon migrations through micro-tags, telemetry, and genetic technologies, scientists also have investigated the mechanisms of salmon homing. What kinds of attachment did salmon have to their rivers (now understood as their "homes"), and how did they manage to find their ways back to them? Since the late nineteenth century, scientists had speculated that odor played a major role in fish orientation (1873, in Quinn 2005), and it remains one of the primary explanations for salmon navigation (see Scheuring 1930; Craigie 1936, in Hasler and Wisby 1951).[7] In 1951, Hasler and Wisby, two biologists at the University of Wisconsin, Madison, formally proposed the "olfactory hypothesis"—that salmon are deeply attracted to the smell of their home stream. According to one account, the olfaction idea came to Hasler when he was walking in an alpine meadow and smelled mosses and columbine near a waterfall he had visited in his youth (Quinn 2005). The odors evoked such strong memories in him that he hypothesized that salmon might form similarly powerful memories of their natal streams (Hasler and Scholz 1983). While Hasler did not believe that salmons' sensual memory connections were identical to those he experienced as a human, he did think they might be a form of imprinting, in this case, a process of olfactory learning during an especially sensitive period, during the smolt stage (Quinn 2005, 91). A special and limited imprinting period, Hasler argued, was necessary to explain salmon's intense drive to return to their natal river.

Such ideas further naturalized (and sometimes also romanticized) these fishes' connection to place. They depicted salmon as organisms that had a profound and fundamental connection to the sites of their birth, an easy analogy for human attachment to homeland that was promoted within many nationalist discourses. Hasler and Wisby sought to experimentally test their olfactory hypothesis and the strength of salmon attachments to natal place. To do so, they exposed juvenile hatchery smolts to a novel synthetic chemical, morpholine, and then scented a specific area of a stream with that same chemical.[8] The chemically imprinted fish returned to that area, while nonimprinted fish, used as a control, did not home in to the chemically scented zone (Hasler and Scholz 1983).

Nordeng, a Norwegian biologist, sought to build on this idea that scent mattered while adding a key twist: he suggested that the key scents to which salmon homed might be produced by the fish themselves. Based upon studies of Arctic char, trout, and Atlantic salmon in the river Salangen in Norway, Nordeng proposed that homeward navigation was made possible by population-specific pheromone trails. In his theory, juvenile fish released pheromones through their skin mucus, creating a trail of pheromones down their home river during their seaward migration (Nordeng 1977, 1971). This trail, he proposed, indicated the way home for their upstream migrating relatives (Nordeng 1977). Nordeng's argument rested on experiments with char that showed that fish sometimes returned to their natal river, even if they had been removed from it as eggs and reared in a hatchery on another river. The fishes' very limited exposure to their natal river led Nordeng to conclude that the fish were instinctively attracted to the smell of their kin, rather than to that of a river they had hardly smelled. Such research depicted salmon as attracted to a home that was defined not only through physical location but also through relations of kinship and family.[9]

While narratives of salmon returns commonly carried romanticized notions of home, the experiments that established ties between particular salmon and particular places were often violent. In a 1971 review publication, one biologist extensively lists early and mid-twentieth-century experiments on fish homing and the methods used (Stasko 1971). From this text, we can see that the methods used to impair senses of smell or vision within homing experiments were often mutilating: cutting the olfactory nerves, cauterization of the olfactory rosette or plugging of the nares with substances, such as Vaseline, waxy material, latex, cotton wool, or cotton wool soaked in a variety of substances, some of which chemically damaged the sensory surface of the olfactory rosette (Stasko 1971, 15). Technologies of blinding involved cutting the optic nerve, removing the eye, removing the lens, replacing the lens with a mixture of petroleum jelly or carbon black, injecting formalin or benzothorium chloride into the pupil, or placing opaque caps over the eyes. If infections and irritations resulting from the operations were minimized, the loss of eye caps was assumed to be the best technique for blinding the fish (Gunning 1959, in Stasko 1971, 18). We mention these methods to further stress that the notion of rivers as salmon homes was not "natural" or preordained and that homemaking is not always a pleasurable experience for all involved. Rather, the processes of linking fish to place and establishing concepts of fish homing were at once labor intensive for people and painful, disruptive—and even lethal—for the fish enrolled in such experiments.

Home and Nation

The scientific practices described in this chapter have had major implications for conceptions of salmon belonging and ownership and for the development of a new ontology of the salmon resource (Richardson & Weszkalnys 2014). By unfailingly configuring the river as salmon home, fisheries research has been integral to molding salmon into a state resource. The scientists who have studied salmon homing have almost universally been part of societies that have held particular conceptions of nation-state units and national citizenship, and these assumptions have shaped both their work and popular interpretations of it. As many scholars have argued, European nation-states have typically been imagined as a classic "domus" writ large—a scaled-up version of a patriarchal agricultural family. Like a family that works together to tend its crops and herds on its own farmstead, the nation typically seeks to maintain productivity within and control over its "household." A nation-state's "domestic affairs" are about the management of its homeland, its people, and its natural resources. Through insistence on firm national boundaries, they have invested extensively in projects of enclosure and control (Ween 2015).

Salmon have been swept up in the commingling of state territorial claims and affective notions of national belonging. Scientific languages of salmon homing—with their focus on the natal stream—have mirrored nation-state notions of citizens as people rooted in their soils. When salmon are linked to a particular river within the bounds of a nation-state, they are configured, from the perspective of that country, as "our" salmon. Scientific descriptions of "homing" serve to firmly attach salmon to rivers and watersheds that are already part of national imaginaries and subject to state territorial claims.

The ability to pinpoint where a fish goes at sea has extended such state territorial imaginaries to ocean worlds and produced a new understanding of seascapes as *dividable* and *bounded*. The nationalization and ownership of fish has been an integral part of the transformation of the ocean itself into a propertied space (Gad & Lauritsen 2009). Echoing Foucault, Pálsson has coined this conversion of the ocean into nationally divided zones in which fish are made into measurable and controllable populations "the birth of the aquarium" (1989, see also Johnsen et al. 2009).[10] Today, as Holm et al. (2009) have pointed out, various international legislation (such as the adoption of the UN Convention on the Law of the Sea in 1983 and the international acceptance of the two-hundred-mile Exclusive Economic Zones for Coastal States), have nationalized more than 95 percent of fisheries resources: Grotius's *Mare Liberum* has been replaced by *Mare Governabilicus* (Caddy & Cochrane 2001; Apostle et al. 2002; Johnsen et al. 2009). Our point here is that fisheries management—drawing on fisheries science—has made fish into countable, commodifiable, and own-able entities (Pálsson 1998, 223; Ween 2012b; Asdal 2011, 2008).

Today, the management of ocean-based fisheries is largely structured around the idea that countries "own" the salmon that spawn in their rivers and that they thus should have the exclusive right to harvest them. It is the history of fish tagging and homing science that has made such logics possible—and it is to these same sciences that managers increasingly turn to address the management disputes that such logics produce.

On the northernmost coast in Norway, for example, a recent Norwegian-Finnish-Russian collaborative study, "Sea Salmon Fisheries: Resources and Potential" (Kolarctic ENPI CBC 2008), was initiated in response to Russian claims of ownership to the salmon that swim along the Norwegian northernmost coast. Fish captured through trial fisheries on the coast close to the Russian border were genetically tested. Results from this testing provided evidence that in some areas, at some points in the season, more than 30 percent of the salmon caught in these Norwegian waters homed to Russian rivers. In a later study, the proportion rose to 50 percent (Svenning et al. 2008; Svenning et al. 2015).

Based on this data, the joint Norwegian-Russian project sought to identify the movements of the different salmon stocks, so that they could make what are called "stock specific migratory models" that would predict where fish from specific rivers would be at different times (Svenning et al. 2015). The international scientific team hypothesized that by using such models, it could become possible to ensure that Norwegian coastal salmon fishermen fish only the Norwegian-origin salmon without catching the Russian salmon that are seen as belonging to the Russian state (Svenning et al. 2015). In other words, they hope that mapped fish behavior can make it possible for transnational fishing agreements to design time-space regulations for fishermen—policies that reinforce the understanding that fishermen have the "right" to fish their "own" nationalized stocks and not those assigned to others (Pálsson 1998; Grey 1998; Symes & Phillipson 2001). Such management models—supported by research on salmon migration—have already been implemented along the Pacific Coast borders between the United States and Canada (see U.S.-Canadian Pacific Salmon Treaty 1985, including revisions in 1999 and 2009; see also Wadawitz 2015).

The possibilities of separating out nationalized salmon stocks and the conceptual justification for doing so are the cumulative product of three hundred years of salmon homing science, from the eighteenth-century clergymen who made the first guesses about fish migrations to the most recent genetic studies. The knowledges of salmon homing produced through increasingly sophisticated tagging, tracking, and identification procedures have been essential to the development of nation-state claims to salmon. Domesticating wild salmon, that is, making them national commodities, has not been easy. Because the oceanic routes of salmon are not directly observable by either researcher or bureaucrat, establishing translocal ownership—by rendering invisible routes visible—has been a difficult, expensive, and lengthy process. In this light, the domestication of wild salmon—the process of tying them to nation-states—is a remarkable and surprising success.

As we have shown, the first achievement of early fisheries scientists was to develop the concept of rivers as salmon "homes." Then, as biologists produced tagging and genetic technologies that allowed them to trace ocean migratory routes and identify the origins of individual salmon, possibilities opened for states to make ownership claims over fish far beyond their territorial boundaries. In this way, the many small tools of fisheries science (Asdal 2015) have come together to produce nationalized salmon, by making them known and hence available to state power (Latour 1987, 218, 1999). As the introduction of regulatory measures and other bureaucratic tools suggest, existing surveillance techniques contribute to a bureaucratic perception of a scientific ability to monitor each salmon stock. Moreover, such precise monitoring opens opportunities for differentiated regulations of particular salmon stocks, whereby authorities can map and control populations and fisheries on each stock (Ween 2012).

Conclusion

This chapter encourages a broader definition of domestication and thinking beyond the literal enclosure and control of animals to allow the concept of domestication to include the making of contained national resources, as in Pálsson's aquarium. As the introduction to this volume points out, to consider more "marginal" domestications—cases that would not be classified as such under strict definitions of the term—makes it possible for us to ask how human relations with not only farmed salmon but also wild salmon might be productively analyzed through a "domestication" framework.

Our chapter benefits from broadening the definition of domestication even as it demands continued attention to practices of enclosure and control. To this end, our work resonates with Tsing's chapter 11, wherein she discourages the rethinking of domestication as a synonym for all human-animal entanglements. Tsing urges us to continue to use it as a critical term for examining relations between bodies, power, ownership, and state forms. For her, domestication is a conceptual tool that sheds light on multiscalar projects of enclosure and accumulation, from family structures to state formations and capitalist enterprises. Our attention to the domestication of the wild salmon reminds us of the importance of one of Tsing's key points: that while domestication may be generative, it may not be generative of just and equitable worlds. The domestication of wild salmon in the North Atlantic has indeed been violent and alienating. The violence has not been limited to the mutilation of research fish but has involved the silencing of other modes of conceptualizing and relating to salmon, as well as claims to them that do not begin with European notions of home (Latour 1999; Tsing 2005). Early in the chapter, we discussed Sámi Veide ways of thinking salmon through notions of return rather than home. As home-based ways of doing salmon have become politically dominant, they have pushed aside other modes of relating to these fish (for similar issues, see Ingold 2000; Blaser 2009; Ween 2012a). Along the Norwegian coast, the Norwegian state has sought to replace Sámi modes of caring for and relating to lands, rivers, and salmon with its technoscientific modes of management and ownership (Ween 2012a).

It is important to remember that there is nothing "natural" about seeing salmon as national citizens under the rule of nation-states. As we have shown, these are ideas that are made and maintained through sets of practices (Law 1989). We have traced the domestication of salmon to the state with the intent of showing that their enclosure within rigid notions of European-style ownership was highly contingent and dependent on certain techniques and knowledges generated by salmon homing science. By seeing such histories as contingent, we hope to open a space for asking questions about how wild salmon and others might become less domesticated or subject to less rigid forms of state ownership, management, and control.

NOTES

1. See, for example, the work of the North Atlantic Salmon Conservation Organization, http://www.nasco.int, or the U.S. and Canadian Pacific Salmon Treaty, http://www.psc.org/about-us/history-purpose/pacific-salmon-treaty/, or, finally, the Pacific Salmon and the North Pacific Anadromous Fish Commission, http://www.dfo-mpo.gc.ca/international/media/bk_pacificsal-eng.htm.

2. The term "straying" casts salmon who do not return home as deviant, even though those fish could just as easily be described as colonizing (as is the case for many other species).

3. These connections are strongly embedded in many European terms for nation, such as *Vaterland* (German) and *la patrie* (French).

4. See http://www.coml.org/pacific-ocean-shelf-tracking-post/.

5. See http://www.nasco.int/sas/salsea.htm.

6. Another technology that allows for the collection of data on salmon routes is the pop-up tag. Pop-up tags record a variety of data about a fish's movements before they pop off the fish and float up to the ocean's surface, where they send out a satellite signal and transmit the information they have collected. These tags provide substantial information about the routes fish take at sea, as well as about the locations where fish from different rivers feed before returning to spawn (Svenning and Prusov 2011).

7. Later research has indicated that salmon may also use the sun and an internal magnetic compass in addition to odor.

8. Morpholine and betaphenylethyl alcohol (PEA).

9. Nordeng's theories remain controversial, and the role of pheromones in salmon homing continues to be explored (Bett and Hinch 2015).

10. The fish that these texts describe were generally migratory species other than salmon.

REFERENCES

Anderson, David G. 2014. "Cultures of Reciprocity and Cultures of Control in the Circumpolar North." *Journal of Northern Studies* 9, no. 2: 11–27.

Apostle, Richard B., Bonnie J. McCay, and Knut H. Mikalsen. 2002. *Enclosing the Commons*. St. Johns: ISER Books.
Asdal, Kristin. 2008. "On Politics and the Little Tools of Democracy: A Down-to-Earth Approach." *Distinktion* 9, no. 1: 11–26.
Asdal, Kristin. 2011. *Politikkens natur: Naturens politikk*. Oslo: Universitetsforlaget.
Asdal, Kristin. 2015. "Enacting Values from the Sea: On Innovation Devices, Value-Practices and the Modification of Markets and Bodies in Aquaculture." In *Value Practices in the Life Sciences and Medicine*, ed. I. Dussauge, C.-F. Helgesson and F. Lee, chapter 9. Oxford: Oxford University Press.
Berthold, Peter. 1993. *Bird Migration: A General Survey*. Oxford: Oxford University Press.
Blaser, Mario. 2009. "The Threat of the Ymo: The Political Ontology of a Sustainable Hunting Program." *American Anthropologist* 111, no. 1:10–20.
Caddy, John F., and Kevern L. Cochrane. 2001. "A Review of Fisheries Management Past and Present and Some Future Perspectives for the Third Millennium." *Ocean and Coastal Management* 44, no. 9-10: 653–82.
Clutton-Brook, Juliet. 1999. *A Natural History of Domesticated Mammals* (2nd ed.). Cambridge: Cambridge University Press.
Craigie, E. Horne. 1926. "A Preliminary Experiment upon the Relation of the Olfactory Sense to the Migration of Sockeye Salmon." *Transactions of the Royal Society of Canada* 5: 215–24.
de Laet, Marianne, and Annemarie Mol. 2000. "The Zimbabwe Bush Pump: Mechanics of a Fluid Technology." *Social Studies of Science* 30: 225–63.
Dingle, Hugh. 1996. *Migration: The Biology of Life on the Move*. Oxford: Oxford University Press.
Foerster, E. 1954. "On the Relation of Adult Sockeye Salmon (*Oncorhynchus nerka*) Returns to Known Smolt Seaward Migrations." *Journal of the Fisheries Research Board of Canada* 11, no. 4: 339–50.
Foerster, E. 1968. *The Sockeye Salmon, Oncorhynchus nerka*. Ottowa: Fisheries Research Board of Canada.
Gad, Christopher, and Peter Lauritsen. 2009. "Situated Surveillance: An Ethnographic Study of Fisheries Inspection in Denmark." *Surveillance and Society* 7, no. 1: 49–57.
Goody, James. 1977. *The Domestication of the Savage Mind*. Cambridge: University of Cambridge Press.
Grey, Tim S. 1998. *The Politics of Fishing*. New York: Springer.
Gunning, Gerald E. 1959. "The Sensory Basis for Homing in the Longear Sunfish." *Invest: Indiana Lakes Streams* 5: 103–30.
Harrt, Allan C. 1966. "Migrations of Salmon in the North Pacific Ocean and Bering Sea as Determined by Seining and Tagging, 1959–1960." *International North Pacific Fisheries Commission Bulletin* 19: 141.
Hasler, Arthur D., Edward S. Gardella, Ross M. Horrall, and H. Frances Henderson. 1969. "Open Water Orientation of White Bass as Determined by Ultrasonic Tracking Methods." *Journal of the Fisheries Research Board of Canada* 26: 2173–92.

Hasler, Arthur D., R. M. Horrall, Warren J. Wisby, and W. Braemer. 1958. "Sun-Orientation and Homing in Fishes." *Limnology and Oceanography* 3: 353–61.
Hasler, Arthur D., Allan T. Scholtz, and Robert W. Goy. 1983. *Olfactory Imprinting and Homing in Salmon: Investigations into the Mechanism of the Imprinting Process.* Berlin: Springer Verlag.
Hasler, Arthur D., and Warren J. Wisby. 1951. "The Discrimination of Stream Odour by Fishes and Its Relation to Parent Stream Behavior." *American Naturalist* 85, no. 823: 223–31.
Hasler, Arthur D., and Warren J. Wisby. 1958. "The Return of Displaced Largemouth Bass and Green Sunfish to a Home Area." *Ecology* 39: 289–93.
Hoare, Ben. 2009. *Animal Migration: Remarkable Journeys in the Wild.* Berkeley: University of California Press.
Ingold, Tim. 2000. *Perceptions of the Environment. Essays on Livelihood, Dwelling and Skill.* London: Routledge.
Johnsen, Jan P., Petter Holm, Peter Sinclair, and Dean Bavington. "The Cyborgization of the Fisheries: On Attempts to Make Fisheries Management Possible." *MAST* 7, no. 2: 9–34.
Jefferts, Keith B., Peter K. Bergman, and Hugh F. Fiscus. 1963. "A Code-Wire Identification System for Macro Organisms." *Nature* 198: 460–62.
Jensen, Jens M. 1968. "Report on Recaptures in Greenland Waters of Salmon Tagged in Rivers in America and Europe and of Recaptures from Tagging Experiments in Greenland." *ICNAF Redbook*, part III: 81–87.
Latour, Bruno. 1987. *Science in Action: How to Follow Scientists and Engineers through Society.* Cambridge, MA: Harvard University Press.
Latour, Bruno. 1990. "Drawing Things Together." In *Representation in Scientific Practice*, ed. Michael Lynch and Steve Woolgar. Cambridge, MA: MIT Press.
Latour, Bruno. 1999. *Pandora's Hope: Essays on the Reality of Science Studies.* Cambridge, MA: Harvard University Press.
Law, John. 1986. "On the Methods of Long Distance Control: Vessels, Navigation, and the Portuguese Route to India." In *Power, Action and Belief: A New Sociology of Knowledge?*, ed. John Law, 234–63. Sociological Review Monograph 32, Henley: Routledge.
Law, John. 2002. "Objects and Spaces." *Theory, Culture and Society* 19: 5–6, 91–105.
Law, John, and Annemarie Mol. 2001. "Situating Technoscience: An Inquiry into Spatialities." *Society and Space* 19: 609–21.
Lien, Marianne. 2015. *Becoming Salmon: Aquaculture and the Domestication of a Fish.* Berkeley: University of California Press.
Lohman, Kenneth J., Catherine M. F. Lohman, and Courtney S. Endres. 2008. "The Sensory Ecology of Ocean Navigation." *Journal of Experimental Biology* 211: 1719–28.
Matthews, G. V. T. 1968. *Bird Navigation.* Cambridge: Cambridge University Press.
McKeown, Brian A. 1984. *Fish Migration.* Beckenham, UK: Croom Helm Ltd.
Miller, Richard B. 1954. "Movements of Cutthroat Trout after Different Periods of

Retention Upstream and Downstream from Their Homes." *Journal of the Fisheries Research Board of Canada* 11: 550–58.

Moser, Ingunn, and John Law. 2007. "Good Passages, Bad Passages." In *Technoscience: The Politics of Interventions*, ed. Kristin Asdal, Brita Brenna, and Ingunn Moser. Oslo: Unipub.

Nordeng, Hans. 1971. "Is the Local Orientation of Anadromous Fishes Determined by Pheromones?" *Nature* 233: 411–13.

Nordeng, Hans. 1977. "A Pheromone Hypothesis for Homeward Migration in Anadromous Salmonids." *Oikos* 28, no. 2/3: 155–59.

Nordeng, Hans, and Per Bratland. 2006. "Homing Experiments with Parr, Smolt and Residents of Anadromous Arctic Char *Salvelinus alpinus* and Brown Trout *Salmo trutta:* Transplantation between Neighbouring River Systems." *Ecology of Freshwater Fish* 15, no. 4: 488–99.

Pálsson, Gisli. 1998. "The Birth of the Aquarium: The Political Ecology of Icelandic Fishing." In *The Politics of Fishing*, ed. T. S. Grey, 209–27. New York: Springer.

Quinn, Thomas P. 2005. *The Behaviour and Ecology of Pacific Salmon and Trout.* Seattle: University of Washington Press.

Quinn, Thomas P., and Kurt Fresh. 1984. "Homing and Straying in Chinook Salmon (*Oncorhynchus tshawytscha*) from Cowlitz River Hatchery, Washington." *Canadian Journal of Fisheries and Aquatic Sciences* 41, no. 7: 1078–82.

Stasko, Aivars B. 1971. "Review of Field Studies on Fish Orientation." *Annals New York Academy of Sciences.*

Svenning, Martin A., Vidar Wennevik, Sergei Prusov, Eero Niemelä, and Juha-Pekka Vähä. 2008. *Sjølaksefiske i Finnmark: Ressurs og potensial. Del II. Genetisk opphavet hos atlantisk laks (salmo salar) anga av sjølaksefiskere langs kysten av Finnmark sommeren og høsten 2008.* Rapport fra Norsk institutt for naturforskning (NINA), Havforskningsinstituttet (IMR), Det finske vilt og fiskeriforskningsinstituttet (RKTL), N.M. Knipovitsj polarvitenskapelige forskningsinstitutt for sjøfiskehusholdning og oseanografi (PRINRO-Murmansk) og Universitetet i Turku (UTU).

Svenning, Martin A., and Sergei Prusov. 2011. "Atlantic Salmon. In *The Barents Sea: Ecosystem, Resources, Management; Half a Century of Russian-Norwegian Cooperation*, ed. T. Jakobsen and V. K. Ozhigin. Trondheim, Norway: Tapir Akademisk.

Svenning, Martin Andreas, Morten Falkegård, P. Fauchald, N. Yoccoz, Eero Niemelä, J-A Vähä, M. Ozerov, Vidar Wennevik, and S. Prusov. 2015. "Region and Stock-specific Catch and Migration Models of Barents Sea Salmon." Rapport fra Norsk institutt for naturforskning (NINA), Havforskningsinstituttet (IMR), Det finske vilt og fiskeriforskningsinstituttet (RKTL), N. M. Knipovitsj polarvitenskapelige forskningsinstitutt for sjøfiskehusholdning og oseanografi (PRINRO-Murmansk) og Universitetet i Turku (UTU).

Symes, David, and Jeremy Phillipson. 2001. *Inshore Fisheries Management.* Dordrecht: Springer Books.

Taylor, Joseph. B. 1999. *Making Salmon: An Environmental History of the Northwest Fisheries Crisis.* Seattle: University of Washington Press.

Tsing, Anna. 2005. *Friction: An Ethnography of Global Connection*. Princeton: Princeton University Press.
Wadewitz, Lissa. K. 2015. *The Nature of Borders: Salmon, Boundaries, and Bandits on the Salish Sea*. Seattle: University of Washington Press.
Ween, Gro B. 2012a. "Resisting the Imminent Death of Wild Salmon: Local Knowledge of Tana Fishermen in Arctic Norway." In *Fishing People of the North; Cultures, Economies, and Management Responding to Change*, ed. Jahn Petter Johnsen, Courtney Carothers, Keith R. Criddle, Catherine P. Chambers, Paula J. Cullenberg, James A. Fall, Amber H. Himes-Cornell, Nicole S. Kimball, Charles R. Menzies, and Emilie S. Springer, 153–71. Fairbanks: Alaska Sea Grant.
Ween, Gro B. 2012b. "Performing Indigeneity in Human-Animal Relations." In *Eco-global Crimes: Contemporary and Future Challenges*, ed. Rune Ellefsen, Guri Larsen, and Ragnhild Sollund. Farnham, UK: Ashgate.
Ween, Gro B. 2015. "Kevo subarktisk forskningsstasjon: Villmark, arkitektur og kjønn i et vitenskapelig hjem i finsk Lappland." *Tidsskrift for kjønnsforskning* 39, no. 2: 130–49.
Winn, Howard E., Michael Salmon, and Nicholas Roberts. 1964. "Sun-compass Orientation by Parrot Fishes." *Tierpsychol* 21: 798–812.
Wisby, Warren J., and Arthur D. Hasler. 1954. "Effect of Olfactory Occlusion on Migrating Silver Salmon (*O. kisutch*)." *Journal of the Fisheries Research Board of Canada*. 11: 472–78.

10

WILDERNESS THROUGH DOMESTICATION

Trout, Colonialism, and Capitalism in South Africa · *Knut G. Nustad*

Anthropologists and political ecologists have for a long time problematized the relationship between what used to be called nature and society, the wild and the domestic. Colonialism both re-created the English countryside as wilderness and established national parks throughout the colonies as fenced-off celebrations of precolonial wilderness (Cronon 1996; Neumann 1998; Olwig 2002). As this literature reminds us, global colonial and capitalist history and natural history need to be thought about together. Many scholars, from very different academic traditions, have attempted to dissolve this division. The topic of this volume, domestication, is in itself a prime example of the need for thinking about natural and global history together. Writers from Engels (1958) to Tsing (this volume) have underlined how the processes that we call "domestication" have been integral to the division of labor, private property, and state formation. But the environmental challenges we now face necessitate thinking that goes beyond merely pointing out the close connection between the natural and the social. Not only is there a close connection between natural and political history, but these histories overlap to such a degree that we can no longer conceive of either of these terms without the other. There is no longer any outside to either of them. We cannot think of the political without the natural, and vice versa.

Here, I want to explore this problematic empirically by looking at the introduction of trout to South Africa and the shifting debates around its status. The colonial introduction of trout is interesting in this regard, because the practice has always muddled the divide between global political history and natural history. Trout were introduced by the British with the aim of making South African river landscapes more like those of the English countryside at home. What they sought to create was contrived yet, at the same time, also a form of wilderness—albeit an English wilderness translocated to South Africa. In this sense, then, the introduction of trout, together with the introduction of other species, was part of an attempt by the colonists to domesticate the landscape.

A project that began with the domestication of landscapes by humans resulted in trout finding a home for themselves in South Africa. They now reproduce successfully in many rivers. Their ability to colonize South African rivers, however, has led to accusations of invasiveness. Here, rather than going into the alien species versus the social construction of alienness debate, I would like to explore what an introduced species such as trout can tell us about living in a world that has been fundamentally transformed by humans. In other words, I want to treat trout in South Africa not as a symbol of the colonial wreckage of nature but as an unusual and unexpected form of domestication, with its own internal contradictions. This is a form of domestication in which what is sought is a human-animal relationship halfway between the domus and the wild. Trout were introduced both to reshape environments to resemble "home" and to create a specific form of nature, if not wilderness, separate from human society. South African trout, then, are a product of colonial expansion. They are a neoliberal, postcolonial subject, as well as a product of a particular form of antipodean and colonial ordering of nature and are perhaps good to think with when imagining living in an environment profoundly shaped by humans. The case of trout in South Africa, I propose, also helps us to rethink the division between the wild and the domesticated. Rather than dissolving this distinction, I argue that trout in South Africa help us to realize how these categories are also contained within each other.

The Introduction of Trout: Disturbing the Nature-Society Divide

As has been well documented, European colonial expansion involved the movement not just of people but also of animals and plants (see, for example, Crosby 1973; DeJohn Anderson 2006; Grove 1996; Lever 1992; Lien and Davison 2010; Lien 2005, 2015). The landscapes of the colonies were often seen as lacking in plants and animals compared to the nature back home. W. Wardlaw Thompson wrote, "The colonist, especially if of British blood, seems unable to settle down in a new land until many of the animals and plants that ministered to his pleasure or profit in the homeland have followed him" (in Brown 2013, 31–32). The same was true of fish, especially trout.

Attempts to introduce trout in South Africa began in the 1860s, following their successful introduction from Britain to Australia, New Zealand, and Tasmania. In 1867, the Cape Colonial Parliament passed a bill "encouraging the introduction into the waters of this colony of fishes not native to such waters" (Curtis 2005, 36). Having worked out how to transport ova over such long distances, it was thought that the relatively short shipment to South Africa would, in comparison, be easier. Not so. It proved to be a difficult task, and the first successful stocking of a river did not take place until 1890, when the Mooi River was stocked with five hundred fry on 2 May, followed by the Bushman's River on 7 May and Umgeni on 10 May (Brown 2013, 37). Draper (2003) claims that the Bushman's and Umgeni Rivers were the first to be stocked and to sustain self-reproducing populations of trout.

Sydney Hey, in the South African classic *The Rapture of the River* describes this early process of acclimatization (Hey [1957] 2006). He was given the responsibility of the Inland Fisheries Review, a commission that involved traveling throughout the union to assess the conditions of its fisheries and make recommendations about stocking. According to Hey, rivers were stocked with both rainbow trout and brown trout, but he and other writers claim that the rainbows tended to displace the browns. Hey, for one, did not mind this, as he saw the rainbow as providing the better sport. This displacement occurred in two of the more famous rivers in South Africa, the Eerste and the Lourens, both of which were stocked with brown trout in the 1890s but are now rainbow trout rivers. As Duncan Brown has pointed out, the rainbow trout must be the ultimate postcolonial subject: native to the Pacific coast of North America and translocated to South Africa from Britain (Brown 2013, 38, see also Halverson 2010).

Once introduced, the acclimatization of trout was a huge success. This was partly due to the trout's ability to adapt to the new climate but also to the elite political support they enjoyed (Curtis 2005). Trout were protected by the Crown, and annual reports of the success or otherwise of their acclimatization were delivered in Parliament. Neither was there, as Curtis also points out, any opposition from nature conservation groups. On the contrary, the provincial nature conservation bodies prioritized the conservation and growth of trout stocks as one of their main duties. Draper (2003) argues that the introduction of trout could never have taken place without the active backing of the state and of elite interests. When John Geddesed-Page was appointed to the Natal Parks Board with a special responsibility for fisheries, the following incident occurred:

> I had settled in comfortably and was picking up the requirements of the new job from "The Colonel" and from the Fisheries Research Officer, Bob Crass, when the Judge (who was a member of the Board at that time, and who shall remain nameless!) called me to his Chambers. Imagine! I was somewhat overawed. Well, I was duly informed that he was the Chairman of the Trout Fishing Liaison Committee of the Natal Parks Board and he then spent a monologuish twenty minutes telling me exactly how I should run the Board's Inland Fisheries Department in general and how I should attend to the best interests of Natal's trout fishermen in particular. (In Draper 2003, 69)

The special protection accorded to trout went unquestioned until 2003, when Khulani Mkhize was appointed director of the Natal Parks Board (Draper 2003, 69).

In early accounts written by people who introduced trout to South African waters, there is a clear distinction made between introduced and indigenous species. But, unlike contemporary writings, this distinction is not mapped onto a nature versus society divide. Rather, nature is seen as being improved by the stocking of rivers with trout (see Franklin 2011; Lever 1992; Lien 2005, 2011). What is interesting, though, is that the nature-society divide is made real through narratives that are at once similar to and different from those found in contemporary writings. For example, a nature-society distinction that resembles contemporary distinctions is invoked in discussions of soil erosion and pollution. Hey was obsessed with the negative consequences of soil erosion for rivers and for fish and, when farmers plowed fields too close to the river banks, castigated them for their unsustainable practices and ignorance. In describing this issue in *The Rapture of the River*, he departs from the usual style of the book, which, despite its poetic title, is written in a somewhat dull and prosaic fashion:

> I visited Port St. Johns again some 34 years later and although the scenery was as strikingly beautiful as ever, its fame as a fishing resort had long since departed, due solely to the millions of tons of valuable soil that is now annually washed down from this river's vast catchment area. The deplorable state of the Umzimvubu estuary today, compared to what it was forty years ago, is an example of the evils of soil erosion, which is entirely due to man's bad soil management. In a book which I read recently, "*Tibetan Marches*," Dr. Andre Migot, writes: "There is in point of fact only one animal on our planet which deserves the French epithet *sauvage*, and that is *Homo sapiens* himself." With this remark I fully agree, for not only has man destroyed many of this planet's fauna, to the point of extinction in some places, but also much of its flora, and is now stupidly engaged in dissipating the living soil on which his very existence depends. (Hey 2006, 34; emphasis in original)

Later, in the same chapter, he writes: "When I see a river heavily charged with silt flowing inexorably to the ocean, it reminds me painfully of some inarticulate animal slowly bleeding to death, the sea absorbing the life blood of the earth as irretrievably as the soil absorbs the blood of the dying animal" (Hey 2006, 35).

THIS SENTIMENT IS FAMILIAR TODAY, with one huge exception: Hey's main concern was to stock alien species into the rivers he cared for so much. In the writings of the people who helped introduce trout into South African waters, then, we see a somewhat familiar nature-society binary, but whereas human society is seen as destructive to environments in ways that ring familiar a hundred years later, alien fish stocks in South African waters were viewed, in Hey's day, as being a part of "nature." Trout thrive in clean, unpolluted water and are therefore seen as belonging to the nature side of a nature-society divide, even if a clear distinction is made between alien and native fish. The aim of successfully establishing introduced fish in South Africa was clearly similar to that described by Lien (2005) and Franklin (2011) for Australia.

The point I want to note here, however, is that there seems to be no contradiction for these authors between reshaping the landscape through acclimatization and the introduction of plants and animals, on the one hand, and upholding a distinction between nature and destructive human intervention, on the other. The nature-society divide is differently configured. The main difference seems to be that during the colonial era, while a distinction between nature and society was upheld, it did not encompass the idea of an untouched, wholly natural nature. Nature that had not been destroyed by industrial agriculture was cherished. As Brown has pointed out, an important argument for stocking rivers with trout was that they provided an important source of protein for those who were allowed to fish for them (personal communication). Clean, unpolluted rivers provided food as well as good sport. Thus, this version of a nature-society division accommodated trout unproblematically.

The Trout Wars

This was to change in the 1980s. Until 1986, trout were unique in having an apparatus of legislation and policies aimed at their protection. Only fly-fishing was allowed, closed seasons were observed during spawning, waters were stocked, and minimum sizes and bag limits imposed. Trout were protected both as nature and as a resource. But on January 1, 1986, the Cape Department of Nature and Environment Conservation (CDNEC) announced that they would withdraw all legal protection for trout. According to Duncan Brown, this was, in essence, a sensible position: CDNEC did not see it as their main concern to protect an alien species and wanted others to take over the responsibility. This highlights how, toward the end of apartheid, concerns about alien species and indigeneity suddenly appeared on the South African ecological and political agenda, a process that the Comaroffs (2006) have argued must be understood as part of a wider conversation among whites about belonging and alienness during the country's transition to democracy (Lien 2007 has pointed to similar processes in Tasmania; see also Nustad 2015 and Lien and Law 2011 on biomigration as a threat).

The heated debate that followed CDNEC's decision has become known as the "trout wars." It ended in a truce: the CDNEC transferred responsibility for managing trout to societies such as the Cape Piscatorial Society and even helped establish control over important trout waters. But from written accounts of the debate one gets the impression that more was at stake than the future of an introduced species. Brown speculates that the debate evoked a perceived connection between trout fishing and elitism and, in the South African context, divisions between Afrikaners and the English. At the height of the debate, the chief director of the CDNEC, Dr. Johan Nettling, said in a statement that trout anglers were "self-confessed elitists who without blinking or blushing refer to all other freshwater anglers as coarse fishermen" (Brown 2013, 30). Likewise, the South African trout fishing legend Tom Sutcliffe borrowed the rhetoric of the Nationalist Party when he described the antitrout lobby's position as a "total onslaught." For the trout, this did not at first make much of a difference, and the result was that the responsibility for managing trout fisheries was passed from the government to private landowners and fishing clubs.

But in 2004 the debate resurfaced in connection with the adoption of the National Environmental Management: Biodiversity Act (NEM:BA). As has been well documented in alien species debates elsewhere, the debate now took on uneasy overtones of biological and racial purity (Brown 2013, 55). Jim Cambray of the Albany Museum at Rhodes University, for example, likened alien species to pollution in an article titled "Time to mop up the 'alien fish spill'!" and described those who were involved with the rearing and stocking of trout as the "true eco-terrorists" (Cambray 2003). This argument, a position that Paul Waley has referred to as econationalism, is common throughout South African environmental discourse. The natural is defined as ecosystems predating Western colonial expansion, static and unchanging (Nustad 2015). The absurdity of the position is clear when Chambray defines indigenous in a way that seems to exclude the Bantu population of southern Africa: "Those waters in which you cast your fly should hold populations of indigenous species which have evolved with the system over many thousands of years" (in Brown 2013, 56).

Debate around the NEM:BA resulted in the proposal of a zoning system, whereby introduced species such as trout would be allowed only within clearly demarcated boundaries and exterminated outside these. Permission to stock new waters had to be applied for. This was a position that many in the trout lobby were quite happy with, but in 2014 the debate flared up again. The Department of Water and Environmental Affairs (DEA) proposed an amendment to the NEM:BA, listing a total of 352 invasive alien species and how to manage them. Among the species listed were brown trout and rainbow trout. The response from the Federation of South African Fly Fishers (FOSAF) and other interest organizations was twofold: first, they argued that while trout were clearly alien, they should not be treated as invasive. As a hundred-year history has shown, they argued, trout do not tend to spread to new waters, even if they sometime do displace native fish in the waters where they are stocked. Second, they pointed out that listing trout as an invasive alien species would entail policies to contain and destroy them, and that this would ruin the 1.4 billion rand trout industry. The industry comprises around forty hatcheries that produce trout both for food and for stocking rivers and waters for fishing, in addition to trout-based real estate and tourism (du Preez and Hosking 2011; du Preez and Lee 2010). This second strand of argument was, according to some, targeted at a contradiction in the alien species listing: introduced species such as pines, wattles, gums, and vines were not listed, presumably because of their economic importance. By stressing the economic importance of trout and the trout industry, it was hoped that trout would be included in the same category. For the time being, this seems to have worked: when the list of invasive species was finally published, it did not contain trout. According to FOSAF, this is because discussions are still ongoing between the parties.

Wilderness through Domestication

The strategy used by FOSAF opens the possibility of an alternative interpretation that what really was at stake here was the viability of an industry, rather than notions of wilderness and domestication. But I would still argue that these concerns underlay the debate about the future of the trout industry. A debate in the South African journal *Flyfishing* bears this out. In October 1997, it featured the story of a rainbow trout weighing fifteen pounds, two-and-a-half ounces, a record in South Africa, caught in the waters of Trout Hideaway in Mpumalanga. The fisherman, Hansford-Steel, recounted in great detail how he had been stalking the fish for days, waiting for the perfect conditions for catching it. The problem, in many people's eyes, was that this fish was a product of a strategy aimed at producing a trophy: not only had the fish been raised to a great weight in the hatchery—it had, in fact, just been released—but it was also swimming in a private dam so overstocked that it and its fellows had to be sustained by daily pellet feeding. Recrimination rained on Hansford-Steel in subsequent issues, with people asking him how many pellet-fed monsters he needed to boost his ego (Brown 2013, 100). His and Trout Hideaway's response is interesting: they argued that since South African water ecosystems are so different from those of Europe and America, practices such as raising trout in hatcheries and supplementary feeding are necessary in South Africa. All trout in South Africa, they claimed, are artificial in this sense.

Duncan Brown does not agree and claims that one of the things trout fishers value the most when they fish is a degree of wildness. A list, he argues, would look something like this: at the top, wild-spawned river fish, followed by wild-spawned lake fish, again followed by river and lake fish stocked as small fish and let to grow naturally; near the bottom, put-and-take fisheries, where fish are stocked at a size ready for capture but in environments where there is enough food for them to sustain themselves; and at the very bottom, stocked fish in environments where the fish cannot sustain themselves without artificial feeding (Brown 2013, 103).

What is at stake here is perhaps a second process of domestication, one undertaken by the trout. When trout successfully establish themselves in a river, and are able to domesticate it as they have in some South African waters, humans see them either as close to wild or as invasive, depending on their point of view, whereas those that are raised in hatcheries and subsequently fed pellets are domesticated in a manner more like cattle. These can never aspire to be wild.

Little is known about how trout have established themselves in South African waters, apart from the fact that they have been able to do so. Franklin (2011) describes how trout played an active role in the domestication of Tasmanian waters, shifting their feeding habits to terrestrial insects and therefore seeking out more shallow areas closer to the lake shores than their European counterparts. This, in turn, Franklin argues, has led to the evolution of a distinctive type of fly-fishing—one that favors sight fishing in the shallows.

I have not been able to find any comparative information for South Africa. The closest is the doctoral research of a freshwater conservation biologist, Jeremy Shelton, on the impact of trout on indigenous fish and insects in the Cape Fynbos Region. Shelton investigated the prevalence of native fish and invertebrates in rivers with and without trout (Shelton et al. 2014a, 2014b; Shelton 2015). He concluded that trout do seem to displace native fish but, perhaps surprisingly, that there was a higher prevalence of invertebrates in rivers with trout, especially mayflies of the *Baetis* family. This is interesting, because these insects, especially the *Baetis rhodani*, provide some of the most sought-after fishing in Europe, in the United Kingdom, in particular. The flies are small and the presentation has to be perfect for the fish to take the fly. Shelton explains the larger prevalence of insects in trout rivers by the combined facts that trout are more likely to rise to feed on the surface than are native fish and that their main food supply is terrestrial insects rather than the mayflies of the English chalk streams. This, then, seems to be an adaptation on the part of South African trout, but it is a finding contradicted by the earlier writings of Crass (1986, 26), who claims that mayfly nymphs are the main food source for trout.

Domesticating Fish and Masculinities

Part of the outrage against Hansford-Steel, I would suggest, and why it is not considered okay to fish domesticated trout in artificial dams, is that practices such as fishing, in addition to the obvious impact it has on the fish, also create people and subjectivities. Again, Sidney Hey will be my guide. In his book, he describes how his transformation from a boy interested in fishing to a fishing fanatic came with the discovery of trout in the early 1920s.

Having been posted to King William's Town, he met Grant, a Scottish immigrant, and together they took up trout fly-fishing. How he learned this skill is fascinating reading. They ordered fly-fishing tackle from England and spent hours practicing before they each managed to catch a trout. Then, as the season closed, they sought ways of learning more:

> Grant was thorough and capable in all he did; so, when our fishing ended for the season, he set about improving our knowledge of trout and trout fishing, as far as this could be done by means of books and fishing tackle catalogs. In addition to Izaak Walton's "*The Complete Angler*," he bought Sir Edward Grey's "*Fly Fishing*" which, by the way, I consider the best book ever written on trout and sea trout fishing, Sheringham's "*An Open Creel*"—a most charming book—E. M. Tod's "*Wet Fly Fishing*," W. C. Stewart's "*The Practical Angler*," A. Ronald's "*The Fly Fisher's Entomology*," which, although it dealt with insect life in English waters only, taught me quite a lot about our ephemera and nymphal life in general. This knowledge stood me in very good stead in after years. Two other good books he also got were J. C. Mottram's "*Fly Fishing*," which is chock full of useful information, and G. E. M. Skues "*The Way of a Trout With the Fly*." Some of these books dealt only with dry fly fishing, which is an art seldom practiced in South Africa; nevertheless the knowledge we gained from our studies of them was put to practical use whenever the opportunity rose. (Hey 2006, 39; emphasis in original)

They thus turned to the English classics, written about chalk stream fishing in England, reading books from the 1650s (Walton) and the 1920s (Skues) and the debates between Skues and Halford and Marryatt over the appropriateness and indeed the morality of using any other method than upstream dry fly-fishing for a rising trout (for more on the moral aspects, see below). Hey had English roots, and his companion had immigrated from Scotland, so it is not surprising that the two should turn to English literature when they set out to explore fly-fishing for trout.

In addition to being inspired by English literature, they also sought to make South African rivers resemble home by introducing trout. But why trout? The rivers contained many species, among them yellowfish, which today is considered equal to trout in terms of sport. These do not count for Hey, who refers to unstocked waters that might be suitable for trout as "virgin waters." One answer lies in the idea of acclimatization, and one of the first fishing clubs in South Africa was indeed called the acclimatization society.

I would argue, however, that as important as the transformation of the rivers was the transformation of subjectivities. Trout in the rivers made possible a semblance of the relationship between man and fish that had developed for centuries on the English chalk streams. Mary Douglas, in a little-known essay from 2003, "The Gender of the Trout," examines the high day of English chalk-stream fishing. She notes that, in contrast to hunting, which often is imbued with sexual metaphors—for example, the male hunter laying down the female prey—trout were, at the time, referred to as masculine. She sees this as resulting from an attempt to cast the trout as an equal, an adversary who chooses voluntarily to take the fly and engage in combat with the fisherman. In the England of the 1880s, the new professional classes were concerned with protecting their interests, setting standards for proper behavior, and policing social borders. Proper conduct on the trout stream became an important arena for this. The social distinctions enacted there were helped by having a material basis as well: only gentlemen had both the means to join a fishing club with access to a chalk stream and enough time on their hands to spend fly-fishing. Douglas gives the example of how the protagonists in Sir Walter Scott's *St Ronan's Well* (1891) finally decide, after watching him fish, that the stranger they have been observing is a gentleman worth of inclusion in their party: "The squire has seen him casting twelve yards of line with one hand, and says: '. . . the fly fell like a thistledown on the water.' Such skill settled the delicate social question and he shortly received three separate invitations to join the exclusive party" (Douglas 2003, 176).

So how does this very particular English assemblage relate to practices of fly-fishing in South Africa? Turner (1974) argues that one of the effects of metaphors is to bring two concepts together in a process that transforms both. What goes on here could perhaps be labeled a metaphoric materiality: the landscapes of southern England and aristocratic notions of sport, of gentleman and prey, are not only likened and compared to a South African reality but, through concrete material means—fish, flies, rods, lines, embodied skills—the two realities, English fish in South African rivers and human-fish relations originating in the English chalk streams, are juxtaposed, and both humans and animals are transformed. One could well imagine, too, that processes occurring in England toward the end of the nineteenth century also took place in South Africa—the policing of borders and the establishment of rules of proper conduct and of morality. This was certainly the case with hunting (Nustad 2015), and it is certainly the case that, at its inception, fishing for trout with a fly was a practice associated with white South Africans of British descent, although this is now rapidly changing.

Duncan Brown points to this process when he criticizes those who argue for allowing trout to remain in South Africa's rivers on the grounds of the British cultural inheritance that fly-fishing for trout embodies: "Recognising the importance and legitimacy of trout and fly fishing in South Africa is not simply preserving a 'British or American' inheritance: it is acknowledging the incorporation and adaptation of those inheritances into the complexities and enfoldings of what makes up present-day South Africa—whether that be in relation to opera, rap, motor vehicles, cinema, dress, religion, cuisine, ballroom dancing or recreational angling" (2014, 29). What is more, these practices have created not only specific South African human subjectivities but trout subjectivities, as well. Brown continues: "Imported trout have adapted to local conditions, and in many rivers they have established distinctive genetic patterns. So has fly fishing in South Africa—while continuing to draw on and be enriched by historical and contemporary developments elsewhere—developed an identity far removed from the stereotypical elitist 'tweed and deerstalker' or 'pioneer American backwoodsman' associations still conjured up by some authors in their simplistic responses to the trout debate?" (Brown 2013, 29–30).[1]

Domestication and the Wild

What can this particular story tell us about domestication? At first glance it seems to break down the distinction between the wild and the domestic, the natural and the social, in a way similar to that argued by Bruno Latour and others. After all, trout were imported into South Africa by the British to transform what they saw as virgin waters, and this led to a domestication of the South African landscape. But, on second thoughts, the process is more complex: What was transposed to South Africa was not just a species of fish but a human-environment relationship that, in England, was inscribed as being a part of the wilderness, of having elite relations with nature. Trout fishing in England stood in opposition to industrial pollution and the destruction of landscapes. A second process of domestication, that of South African rivers by trout, was thus completely entangled with the first process—with the result that the human-environment relationships were also re-created. This is clear in the way Hey laments the destruction of trout streams by poor farming practices and consequent soil erosion and in the way present-day fishers mourn the industrial pollution and death of the rivers that were once stocked by Hey and his fellows. Encapsulated within the global political process of introducing trout and re-creating colonial masculinities, a distinction between nature and society emerged, albeit one differently configured to how it is today. Rather than drawing the line at native/non-native and invasive, the distinction that emerged followed the line of pollution-pristineness, wherein the latter category included what today is considered an alien species.

Furthermore, within the larger process of the domestication of trout in South African rivers, other ideas of wildness were retained, as we saw in the discussion about the status of the pellet-fed monster in Mpumalanga. This notion of wildness follows a familiar line of thought, one that equates wildness with lack of human intervention. Thus, both the introduced and the domestic trout in South Africa are considered wild if they self-reproduce without human intervention; the degree of wildness decreases as more intervention is needed, with the absolute low point being the fish that cannot sustain itself without being fed.

All this points to a situation wherein the categories of wild and domestic, or nature and society, rather than being dissolved seem to be differently configured. The wild and the domestic are contained within each other, so that domestic fish can be considered as wild, just as wild fish in certain configurations are domestic.[2]

Furthermore, these distinctions matter because wildness, as it is configured within the domestic, aligns with the kind of nature that we sought to protect when these two categories were conceptualized as discrete entities. Thus, in contemporary South Africa, there is a convergence of interest between those seeking unpolluted rivers and those caring for trout. Pollution is not just a technical issue but also an affective one. This was indicated in an editorial from 1947 in the *Piscator*, a Cape Town fishing journal. A. C. Harrison argued that water pollution should not be seen as a problem solvable by industry and technology, as "industry looks on un-polluted water as that usable at the minimum expense for pre-treatment" (in Brown 2013, 21). On the contrary, he argued that anglers and fish scientists do not value rivers for their economic potential: their concern is to keep industry out of the waters as much as possible.

In short, then, what trout in South Africa can tell us about domestication is that the categories of the wild and the domestic, rather than being dissolved, have been reconfigured within each other. As such, they continue to inform how we think about and relate to our environments. Perhaps there is a lesson to be learned from South African trout in this respect. The answer to the challenges posed by our present predicament is not to write off ideas of nature as social constructs nor to think that nature exists only in heavily policed and demarcated parks but to recognize that nature and perhaps even wilderness are emergent possibilities even in landscapes heavily shaped by global colonial and capitalist histories.

NOTES

This chapter is based on published literature, articles, and Internet debates, as well as a few interviews and the author's experience of fly-fishing in Norway and England. The empirical data on which it draws are therefore limited. Thanks to the editors and the reviewers for their insightful comments and probing questions. And also a huge thanks to Duncan Brown for taking time to discuss South African trout with me.

1. As noted above, little has been written about the specific ways in which South African trout and fly-fishing practices have evolved.

2. Swanson (2015) provides a fascinating example of this, showing that the restoration of so-called wild fisheries in Japan is closely linked to the industrial domestication of fish in Chile.

REFERENCES

Brown, D. 2013. *Are Trout South African? Stories of Fish, People and Places.* Johannesburg: Picador Africa.

Cambray, J. A. 2003. "The Global Impact of Alien Trout Species—A Review; with Reference to Their Impact in South Africa." *African Journal of Aquatic Science* 28, no. 1: 61–67.

Comaroff, J., and J. Comaroff. 2006. "Naturing the Nation: Aliens, Apocalypse, and the Postcolonial State." In *Sovereign Bodies: Citizens, Migrants, and States in the Postcolonial World*, ed. T. B. Hansen and F. Stepputat, 120–47. Princeton: Princeton University Press.

Crass, B. 1986. *Trout in South Africa.* Braamfontein, Johannesburg: Macmillan South Africa.

Cronon, W. 1996. "The Trouble with Wilderness: Or, Getting Back to the Wrong Nature." In *Uncommon Ground: Rethinking the Human Place in Nature*, ed. W. Cronon, 69–90. New York: W. W. Norton.

Crosby, A. W. 1973. *The Columbian Exchange: Biological and Cultural Consequences of 1492.* Westport, CT: Greenwood.

Curtis, P. 2005. *Fishing the Margins: A History and Complete Bibliography of Fly Fishing in South Africa.* Johannesburg: Platanna.

DeJohn Anderson, V. 2006. *Creatures of Empire: How Domestic Animals Transformed Early America.* Oxford: Oxford University Press.

Douglas, M. 2003. "The Gender of the Trout." *RES: Anthropology and Aesthetics* 44: 171–80.

Draper, M. 2003. "Going Native? Trout and Settling Identity in a Rainbow Nation." *Historia* 48, no. 1: 55–94.
du Preez, M., and S. G. Hosking. 2011. "The Value of the Trout Fishery at Rhodes, North Eastern Cape, South Africa: A Travel Cost Analysis Using Count Data Models." *Journal of Environmental Planning and Management* 54, no. 2: 267–82.
du Preez, M., and D. E. Lee. 2010. "The Contribution of Trout Fly Fishing to the Economy of Rhodes, North Eastern Cape, South Africa." *Development Southern Africa* 27, no. 2: 241–53.
Engels, F., et al. 1958. "Origin of the Family, Private Property, and the State." In *Selected Works*. Moscow: Foreign Languages Publishing House.
Franklin, A. S. 2011. "Performing Acclimatisation: The Agency of Trout Fishing in Postcolonial Australia." *Ethnos* 76, no. 1: 19–40.
Grove, R. 1996. *Green Imperialism: Colonial Expansion, Tropical Island Edens and the Origins of Environmentalism, 1600–1860*. Cambridge: Cambridge University Press.
Halverson, A. 2010. *An Entirely Synthetic Fish: How Rainbow Trout Beguiled America and Overran the World*. New Haven: Yale University Press.
Hey, S., and T. Sutcliffe. 2006. *The Rapture of the River: The Autobiography of a South African Fisherman*. Johannesburg: Platanna.
Lever, C. 1992. *They Dined on Eland: The Story of the Acclimatisation Societies*. London: Quiller.
Lien, M. E. 2005. "King of Fish or 'Feral Peril': Tasmanian Atlantic Salmon and the Politics of Belonging." *Environment and Planning* 23, no. 5: 659.
Lien, M. E. 2007. "Weeding Tasmanian Bush: Biomigration and Landscape Imagery." In *Holding Worlds Together: Ethnographies of Knowing and Belonging*, ed. M. E. Lien and M. Melhuus, 103–20. New York: Berghahn.
Lien, M. E. 2015. *Becoming Salmon: Aquaculture and the Domestication of Fish*. Berkeley: University of California Press.
Lien, M. E., and A. Davison. 2010. "Roots, Rupture and Remembrance: The Tasmanian Lives of Monterey Pine." *Journal of Material Culture* 15, no. 2: 1–21.
Lien, M. E., and J. Law. 2011. "'Emergent Aliens': On Salmon, Nature, and Their Enactment." *Ethnos* 76, no. 1: 65–87.
Marire, J. 2015. "The Political Economy of South African Trout Fisheries." *Journal of Economic Issues* 49, no. 1: 47–70.
Neumann, R. P. 1998. *Imposing Wilderness: Struggles over Livelihood and Nature Preservation in Africa*. Berkeley: University of California Press.
Nustad, K. G. 2015. *Creating Africas: Struggles over Nature, Conservation and Land*. London: Hurst.
Olwig, K. R. 2002. *Landscape, Nature, and the Body Politic: From Britain's Renaissance to America's New World*. Madison: University of Wisconsin Press.
Shelton, J. 2015. "Trout in the Fynbos: Have They Had Serious Ecological Impacts?" *Flyfishing* (April): 48–52.
Shelton, J. M., et al. 2014a. "Non-native Rainbow Trout Change the Structure of Benthic Communities in Headwater Streams of the Cape Floristic Region, South Africa." *Hydrobiologia* 745, no. 1: 1–15.

Shelton, J. M., et al. 2014b. "Predatory Impact of Non-native Rainbow Trout on Endemic Fish Populations in Headwater Streams in the Cape Floristic Region of South Africa." *Biological Invasions* 17, no. 1: 365–79.
Swanson, H. A. 2015. "Shadow Ecologies of Conservation: Co-production of Salmon Landscapes in Hokkaido, Japan, and Southern Chile." *Geoforum* 61: 101–10.
Turner, Victor. 1974. *Dramas, Fields, and Metaphors: Symbolic Action in Human Society.* Ithaca: Cornell University Press.

11

NINE PROVOCATIONS FOR THE STUDY OF DOMESTICATION · *Anna Lowenhaupt Tsing*

A once-fashionable mystery of domestication ran as follows: all our most important domestic plants and animals were brought into the domus during the few thousand years on each side of ten thousand years ago; afterward, comparatively few new species have entered domestic herds and fields. Of course, scholars said, there are interesting exceptions. But did humans somehow lose the knack for taming wild things? Or did we run out of the most easily tamed species?

This puzzle really impressed me—until I thought about it more carefully. Then I realized that this is only a puzzle to the extent that one carries around a quite peculiar concept, namely, "domestication." With this realization, the puzzle changes: Why are only some kinds of interspecies relations singled out? Why did that standard develop—and what has it meant not just for our imaginations but for our world-building projects? This chapter uses these questions to think again about domestication.[1]

Some guidelines situate my chapter within the work of this volume. First, one goal of this chapter is to prod the study of domestication to pay attention to connections with "civilization" and "home" that have too often been ignored. I bring a feminist alertness to the work of "home" in multispecies narratives; "home" too often sugarcoats captivity. This allows me to look at domestication as peculiar—only one among many ways that humans can relate to other living beings. But these goals require this chapter to use a somewhat different definition of domestication than some others in this volume. Many colleagues would like to open the term "domestication" to all kinds of interspecies relations and to emphasize "mutual dependence, dialogue, and trust" instead of human control (Lien, Swanson, and Ween, this volume). The disadvantage in this, as I see it, is that "domestication" suddenly engulfs all multispecies relations. We lose the ability to see the historical force of forms of interspecies dependence. Worse yet, we run the risk of allowing "civilization" and "home" to inflect all those many relations, whether they belong or not. Refusing such unintentional inflections might allow the term more traction, especially as we grapple with threats to multispecies life on earth. In what follows, then, I discuss domestication within the inequalities and intensities of civilization and home, the better to see the specificity of those interspecies relations. "Domestic" organisms would be those whose species being has changed to a form that survives for human needs. Even where domestic forms were not intentionally bred by humans, they have been maintained by the work of human control.

Second, I aim to explore domestication not just as a narrative but also as a world-making process. When I say, below, that "domestication is a progress concept," I'm not trying to criticize the trope; I'm trying to see how progress becomes inscribed into interspecies relations. I don't think we can unfasten domestication and progress merely by telling a different story; the narratives we know today are figured into landscapes, bodies, and social institutions. It's not that early potato-growers or chicken-raisers were thinking about progress—or "domestication." But state and colonial expansion made use of the materials we now call "domestication," and, over time, their use created a dangerous landscape for multispecies life. In this shadow, one reason to cleave to conventional definitions of domestication is to investigate the threats raised by this program as well as its underexplored alternatives.

Third, rather than contributing a historical or ethnographic example for close study, I make proposals for further research. To open doors, I offer bold provocations, perhaps too bold. My hope is that these might scatter the wild seeds sometimes necessary to revitalize a long-cultivated landscape.[2]

To make it simple to use my theses as a baseline for further discussion, I list them in their baldest form before further exploration:

1. Domestication is a progress concept.
2. Domestication as a lens has made it difficult to see varied interspecies relations involving humans.
3. Viewed within the variety of interspecies arrangements, domestication is a feature of the political economy, not care, intelligence, or survival.
4. Grains captured women, imprisoning us in the domus.
5. Domestic plants followed state expansion and long-distance commerce.
6. Cattle and horses spread war machines, excluding other social arrangements.
7. Refusals of domestication are often problems of political ecology, not human ingenuity or a species' unwillingness to form relations with humans.
8. Alienation and accumulation are threats to life on earth.
9. Unintentional cultivation and domestication-as-rewilding offer hopeful alternatives in imagining multispecies life with humans as a component.

ONE. *Domestication is a progress concept.*

Domestication is not just a matter for small and particular lives, tamed in each other's presence; domestication remade the world. Domestication sets a standard: it is not just a narrative but also a process that sets world history into motion. But how can we assert the world-historical nature of domestication? Why should one intimate relationship change everyone? World history of this sort can only be taken for granted within the framework of progress. Yet "domestication" draws us into to its common-sense world without alerting us to this framing.

We know the word "domus," the hearth as a center of human sociality, as it congealed in Roman times. Yet "domestication" as a feature of human evolution is a much more recent concept. One grounding point is archaeologist V. Gordon Childe's early-twentieth-century coinage of the term "Neolithic Revolution" to refer to the origins of farming. Childe did not invent the storied relation between domestication and progress, which embedded domestication in collective evolutionary time. This embedding was written into the narratives of many eighteenth- and nineteenth-century Enlightenment thinkers; indeed, Childe aimed to wrench that relation away from elite self-congratulation to show the radical potential of human creativity. His contribution was to ground domestication in a materialist archaeology in which ordinary people's accomplishments mattered. Childe was self-conscious about reviving hope in progress in the wake of the senseless violence of World War I. Of his masterwork *Man Makes Himself*, he says: "History may still justify a belief in progress in days of depression" (Childe 1951 [1936], 9).

To this end, he conceptualizes the human past as a series of "revolutions," each building on the last. The first and most important is the Neolithic Revolution, which "gave man control over his own food supply. Man began to plant, cultivate, and improve by selecting edible grasses, roots, and trees. He succeeded in taming and firmly attaching to his person certain species of animals" (59). Only this solid base could give rise to cities—the next revolution.

I first read *Man Makes Himself* in my Introduction to Anthropology course in college. I had never heard of anthropology before, and I loved it, in part through Childe. I savored his attention to the material details of ordinary lives within the sweep of human history. I still feel lucky to have read Childe, who is rarely taught today. Rereading the yellowing pages of my text, I feel the pleasure of my earlier encounter. I had forgotten his rebellious answer to the conundrum with which I began this chapter: If people have become less inventive, Childe argues, it is because we were harassed by armies and subjugated by states. No wonder we lost our early genius!

I mention this pleasure so that you will see that I am not lashing out at domestication, or even refusing its revolutionary stature. My goal is respectful: to figure out how domestication as collective advancement might possibly work. I don't think Childe's revolutions are arrogant stories to replace with humbler stories. But they should not be taken for granted as the natural way of things. If something changed the world, we must ask: why?

Childe's revolutions inspired scholars, many of whom lost his interest in everyday creativity to focus more directly on the forward march of evolution. In the mid-twentieth century, scholars read domestication as an intelligence test for humankind. The editors of a 1969 volume stress learning as the key to domestication: "Mankind took an immensely long time to learn how to gain food by any other means than hunting, fishing, and gathering" (Ucko and Dimbleby 1969, xvii). More recently, however, the intelligence test has been supplanted as researchers take note of domestication's unplanned advances. Helen Leach notes that through most of human history, domestication was unintentional; systematic breeding emerged only in the last three hundred years

(Leach 2007). This insight refocuses domestication studies, not to move beyond world-historical effects but rather to ask just what transformations are being tracked. Leach offers four stages through which humans have increased their mastery over plants and animals. The third stage, of intentional breeding, is key: one might look here for domestication's use in narratives of human control. A consideration of the massive anthropogenic changes of the last three hundred years (Leach's third stage) affirms that this might be one place to track domestication not just in narratives but also in engineering projects. Yet, as Leach argues, these innovations built on a ten-thousand-year history of world-historical changes through domestication.

Leach's broadening of the category "domestication" also speaks to a major contribution of early-twenty-first-century scholars. With "the animal turn" in the humanities and social sciences, scholars have begun to describe the nuances of human-animal relations, as these fall both inside and out of earlier definitions of domestication. The volume in which Leach's article appears is full of thrilling accounts of not-quite-domestication; moreover, other species use humans for their own purposes as much as the other way around (Cassidy and Mullin 2007). Yet Childe's basic progress premises are left intact. Domestication is world-historical; in other words, acts of domestication change the world for everyone.

How might this epoch-shaping work? If I plant a seed, why should everyone else follow? To explore this seriously, I first must show some of the relationships between humans and other species that this framework blocks.

TWO. *Domestication as a lens has made it difficult to see varied interspecies relations involving humans.*

Let me take you to the Meratus Mountains of South Kalimantan, Indonesia, where I lived with shifting cultivators who made small impermanent fields in the tropical rain forest. My "ethnographic present" is the 1980s and 1990s, before much of the forest was felled for corporate uses (Tsing 1995). During that time in the central Meratus Mountains, people lived in small groups next to their fields. After two years, fields began returning to forest, but both old fields and young forests were major sources for livelihood. Meratus relations with animals and plants were shaped within the possibilities of this dynamic ecology. Consider two familiar animals with which we think the domestic: chickens and pigs.

Red jungle fowl (*Gallus gallus*), imagined as the "wild ancestor" of global domestic chickens, are abundant in Meratus rain forests, and they interbreed with chickens raised in Meratus households. Here is what it means to "raise": Meratus set up nesting baskets on bamboo poles under the raised house, where the eggs are safe from some predators. After hatching, small chicks have their food supplemented until they can manage. Then they wander in fields and forests as they like. They find their own food and their own mates and escape predators as best they can. At night, they fly to the roof ridge to roost. There are similarities here with the raising of young hornbills as pets, until they fledge: the hornbills come to visit now and then on their flights across the forest.

The situation is not so different for pigs. The Bornean rain forest is full of bearded pigs (*Sus barbatus*); Meratus say their kept pigs crossbreed with them.[3] Here is what it means to raise a pig: a young pig is set on an old swidden, where there are good things to eat, and it is expected to stay there, because of the abundance of food, until people want to butcher it. There are no fences and no prohibitions. Meanwhile, forest-ranging pigs also like old swiddens and congregate there. Surely, they mingle; humans hunt both there, although forest-ranging pigs, they say, taste better. Is this domestication?

One more animal can open the range for further discussion: the giant honeybee. Giant honeybees (*Apis dorsata*) make exposed combs on the branches of forest trees, and Meratus harvest them for honey (Tsing 2003). The bees are migratory, following forest flowerings. Meratus prepare for them. The bees need light for their dances; Meratus clear high horizontal branches of epiphytes and excess shade. When the season returns, colonies rebuild on those branches so long as people keep them clear and clean.

Would you call these arrangements "domestication"? Regardless of your answer, clearly they have not been world-historical. They have not set a course for progress. Indeed, they distinguish themselves as marginal. These are the kind of arrangements that don't count when we talk world history. Let me add some plant examples that make this problem even more wrenching.

Borneo is an origin point for the cultivation of many tropical fruits, such as durian, mango, mangosteen, rambutan, and cempedak. In this area, most cultivated fruits have close relatives in the forest. Are these fruits domesticated? For some fruits, it is hard to know if a given tree has anything to do with humans. Even the line between cultivation and noncultivation can be difficult to draw. Among Meratus, cultivation often looks like this: People sit on the porch eating fruits, and they toss the seeds out into the brush beyond. Six or seven years later, long after that house has been abandoned, it has a grove of fruit trees coming up from those rubbish piles and just beginning to fruit. Such islands of trees are good places for finding fruit. (This method of spreading fruit is not so different from that employed by other primates in the forest.) Meanwhile, people harvest fruits from many forest trees that have nothing to do with human cultivation. When they are clearing a field, they will spare a good fruit tree, allowing it to continue to produce. Sometimes they will prune or otherwise encourage a fruit tree in the forest just because they like the fruits. Is this domestication?

The concept of domestication obscures many kinds of human relations with other species. Such examples make it clear that domestication is not just one element in the description of human interspecies relations. It is a standard to which not many interspecies relations can rise. One reason we seem so limited in domesticates is that we don't count the many species with which we have other kinds of intimate relations. My foray to the Meratus Mountains also offers the beginning of an explanation: Meratus don't have what counts as domestication because they are at the margins of the political economy that we have allowed to make world history. They are outside of progress. It is to this arena that we must turn to understand domestication as world-making.

THREE. *Viewed within the variety of interspecies arrangements, domestication is a feature of political economy, not care, intelligence, or survival.*

Agronomist Jack Harlan (1992) came up with a brilliant experiment for studying the history of domestication: he harvested wild wheat himself in the "cradle of civilization." What he found was that it was perfectly possible to get a good harvest by foraging with primitive tools. Even in the twentieth century, there was plenty of wild grain. This insight makes a big difference. There may have been lots of local reasons to grow one's own wheat, but human survival was not at stake. Domestication was not necessary to making a living. (Even if some areas experienced want, local conditions do not in themselves add up to world-historical advancement.) Domestication did not create a better quality of food or well-being. And if making a living was as successful without it, it makes no sense to consider domestication an intelligence test. It was just one among many kinds of interspecies arrangements.

The tradition of studying domestication as progress created its own just-so stories: domestication, they said, allowed people to learn to make storage containers, to design tools for harvesting grain, and to settle and build houses. Progress! But research has shown all these to be wrong. In the Near East, sickles were used for foraging wild grasses before cultivated grain (Fuller 2007, 920); Chinese built storage containers for acorns and water chestnuts long before taming rice (915). Foragers in many areas had long-lasting architecture and sedentary settlement (Wilson 2007). The picture I start to get from all this is an ancient world in which people with tame plants and animals had to work out arrangements with other equally prosperous people who did not. The question of who was ahead and who was behind probably never came up.

One of the best accounts I know of such arrangements comes from Ian Hodder's *The Domestication of Europe* (1990), which tracks archaeological findings in Europe for the period that should have included the Neolithic Revolution. In southeast Europe six millennia ago, Hodder finds the *domus*, that is, the decoration and elaboration of the hearth, which he believes is central to the emerging domestication complex. The complex moves to central Europe. Yet in southern Scandinavia six millennia ago, it is missing. This is peculiar because there is plenty of evidence for trade between the settlements of southern Scandinavia and central Europe, and Scandinavians are perfectly capable in all the European technologies. He notices that the trade is selective; Scandinavians are not interested in household decoration, preferring austere interiors. They have perfectly successful foraging economies; they are not concerned with domestication or the domus. Almost a millennium and a half later, these areas start to be drawn into the central European symbolic economy, with its homeworld gender segregations and its crop-and-cattle production. Perhaps the earlier period was a time marked by the successful practice of live and let live.

What allowed the domestication complex so much power, such that others were forced to convert or retreat? Hodder helps us here: the symbolic and material domestication of women was at the center of the growing hegemony of a world-system in the making.

FOUR. *Grains captured women, imprisoning us in the domus.*

Working from his excavations in ancient Turkey through the archaeology of Neolithic Europe, Hodder takes us from domestication as an interspecies relation to the social and symbolic complex centering on the domus: domestication becomes world-historical through the political economies of which it forms a part. From the start, in Hodder's reading, the domus situates women, along with death, at the hearth. Later, he argues, the male symbolism of the outside, the *agrios*, takes over, encapsulating the female hearth. Hodder stresses that his archaeology only allows him to study gender symbolism, not the actual lives of ancient men and women. But it is difficult not to imagine that such symbols aim for the domestication of women. They elaborate women's roles in reproduction and food preparation to the detriment of other abilities, particularly the mobility of foraging. All those figurines, with their vulvas and breasts, emphasize only women's fertility. Even before the encompassment of the domus by the masculine *agrios*, women's mobility is gone, at least in a symbolic register.

It is hard to know whether early Neolithic versions of this complex were geared toward conquest and expansion—in part because archaeologists sensibly focus only on the piled up remains of domus-oriented people to the neglect of scattered foragers. Perhaps there are hints: Hodder argues that European Neolithic settlements became increasingly crowded with technologies of war and defense (163). Surprisingly, wild animal bones increase (164), perhaps suggesting that surrounding foragers have been driven back or incorporated.

What we do know is that, millennia later, the woman-and-hearth complex becomes a key tool of all the projects of domination and conquest known to humankind: the state, world religions, and most recently, nationalism. Almost everywhere, states have made male heads of household their local representatives, with women and children understood as their dependents as much as herds and crops are. There is a surprising similarity of family forms among those we call "peasants," that is, traditional rural cultivators subjugated by states. Among those at state margins, a much wider variety of gender and kinship arrangements still flourish. I think of the sharp contrasts, for example, between the lowland peasants and the Meratus hill people of my fieldwork in Borneo. In the lowlands, women are obsessed with having children because their status depends on it; barren women make pilgrimages to beg for fertility. In the hills, women work hard to keep childbearing late and sparse. They want the mobility, livelihood options, and active sociality that too-much childrearing blocks. Their interpretation influences mine: confinement in the domus may be a benign imprisonment, but it is still unfreedom.

Going back to the Neolithic, it is useful to speculate on the ways that grain domestication transformed women. The high carbohydrate diets of the starch-heavy grains of domestic cereals would have increased the frequency of ovulation. Women seem to have begun reproducing earlier and had more babies. Meanwhile, women's stature decreased due to the change in diet.[4] Biological anthropologists have argued that the death of women in childbearing, rather than being a species problem of our famed big heads, becomes acute only with grain domestication, when nutritionally stressed young women whose pelvises have not reached earlier sizes start bearing young (Wells, DeSilva, and Stock 2012). At the same time, the changes wrought by cereal domestication, such as nonshattering racemes, would have created more food preparation work, probably done by women at home. All this could have created two effects: first, bigger, nonsustainable populations, and, second, more support for expansion activities through the specialization of a service class—that is, women—dedicated to providing for others.

Unthinking analysts often assume human populations just "naturally" expand; but reproduction dynamics are always an effect of the political economy of gender. The domus complex seems likely to have started the now-epic human aspiration to overpopulate the earth. This is a good way to begin unsustainable conquest.

FIVE. *Domestic plants followed state expansion and long-distance commerce.*

States capitalized on and transformed the political economy of domestication. Domestic cereals in permanent fields worked particularly well for the ambitions of states: permanent fields kept subject populations in place, ready to pay taxes and available for corvée labor (Scott 2009). Relatively recent histories can make the point. When Dutch traders first arrived in Java in the seventeenth century, they encountered Javanese kings who ruled through the expansion of wet rice cultivation. Rice paddies were considered the epitome of beauty and order; other landscapes were savage, their residents singled out as criminal (Day 1994). In the nineteenth century, the Dutch took over those Javanese landscapes, and they extended the logic of Javanese kings many times over. One of the key principles of colonial rule in Java was the extension of permanent field cultivation—not only in rice, but also in coffee and sugar, crops the colonists wanted (Elson 1994). Colonial armies enforced this policy, settling the rural populace. With the expansion of colonial rule, the Javanese landscape was transformed from a patchwork of temporary clearings, woodlands, and old forests, with a few valleys of concentrated paddy, into almost sea-to-sea permanent cultivation.

During the epoch of European colonial expansion, since the fifteenth century, capital has come to govern the expansion of agriculture as much as state authority. The story of sugar illustrates. Until quite recently, many sources mistakenly said that sugar originated in India.[5] In fact, sugarcane is a New Guinea domesticate; India is important as the first place that crystallized sugar was manufactured as a commercial product. The confusion between biological and commercial origins exemplifies the problem of knowing domestication. Domestication as progress depends on the power of states and capital, and so it makes a strange sense to identify commercial origins as centers of world-historical domestication, even when they merely used plants and animals from other places.

Commercial sugar traveled as cane to the Mediterranean; Muslims controlled the trade, much to the annoyance of European Christians. In Christian attempts to grow their own cane, the European New World plantation complex was born. Capital came into its own; the profits from sugarcane plantations helped fund—and inspire—the industrial revolution. This kind of spread of domestication is at the heart of what we know as progress; it is easy to see earlier human accomplishments supporting later ones. But two odd characteristics of this civilizational movement seem worth pointing out before we naturalize the process:

First, the "taming" of plants that accompanied commercial spread depended on the *discipline* of plants for economies of scale. Plants were remade as interchangeable units to allow larger and larger management schemes—all the better for extraction of profit by investors. This process of increasing scalability has entered common sense about domestication. But it is best understood in its relation to investment.

Sugar led the way: the planting of clonally identical canes showed what was possible. Irrigation was introduced to make the canes grow and ripen evenly; the plant did not need irrigation in well-watered areas, such as Brazil and Java. Full control was the object.[6] Similarly, perennial cotton on New World plantations was treated *as if it were an annual* (Porcher and Fick 2005). Full control was the object. One could enter these into the annals of domestication, but only by confusing the commercial advantages of scalability with the advancement of humanity (Tsing 2012).

A second odd fact: the process of European expansion *decreased* the number of domesticates on earth. In the earliest period of European conquest, new plants and animals were brought into the fold of European interest. However, once plantation economies became successful, commercial interests ensured that the landscapes of the Global South became covered with the plants and animals of greatest interest to the colonizers. Locally popular domesticates disappeared—and are still disappearing, particularly from Latin America and Africa. Consider again Childe's question about the decreasing rate of domestication. One point to keep in mind is the eradication of those domesticates not in the forefront of elite interests in the spread of plantation agriculture. While a few new organisms have been tamed for agribusiness, the number is much less than those retired from the human stock of companions.[7]

SIX. *Cattle and horses spread war machines, excluding other social arrangements.*

In the last two sections, I have focused on plants; what about animals? Animal studies scholars have paid much attention to the cospecies intimacies in which humans and animals tame each other. But close relations between humans and animals can also spread terror. I won't mention the new diseases that living with animals has spread among humans. I am interested in war machines. The term comes from Deleuze and Guattari (1993), who love the war machine for its masculine fury against the state. Their model is the horse-riding barbarian horde of the Eurasian steppes, which once caused so much trouble in Europe. Yet Deleuze and Guattari's war machine is a freedom opportunity mainly for men. Besides, it seems to me that colonizers backed by states have often used the same tactics: taking advantage of the mobility of horses and cattle to spread their violent claims across the landscape. My war machines, then, confuse state and barbarian conquests to focus on the expansionist possibilities of human relations with horses and cows.

Let me begin, then, with state-backed colonization. Virginia Anderson (2006) has documented how British settlers were aided in their expansion across North America by their cattle and pigs. They let their animals roam in the forests and fields of Native Americans. When Native Americans killed or confiscated one of the animals, settlers set their armies and courts on whole communities for violating private property. The animals, Anderson shows, could push the radius of British settlement far beyond the daily activities of human farmers. Settlers let the animals define colonial territory, forcing native peoples to retreat. This is the kind of war machine cattle can support.

Spanish colonization of the Americas was even more explicit about the use of horses and cattle to extend empire. When the Spanish invaded the Americas, they brought their horses, sheep, and cattle to take the land.[8] Open-range grazing emerged from this history of conquest; it helped to drive back native plants and animals as well as indigenous people. We see the continuing conquest in tropical Latin America today, where forests are still being cleared and exotic forage species planted to drive out indigenous residents, re-creating their landscapes as private ranches (Ficek 2014). The aggressiveness of the ranchers is legendary; across the Amazon, for example, indigenous activists and environmental advocates are threatened and murdered.[9] The masculine fury of the war machine is recognizable here—but, contra Deleuze and Guattari, it uses the power of citizenship, the state, and capital.

I draw my bravery to propose a connection between domestic cattle and war machines from the even bolder proposal of Deborah Bird Rose, who, in *Reports from a Wild Country* (2004), argues that wherever cattle and horses have spread they have brought the "wildness" of colonial settlement. Cattle in Australia, she argues, undo the work of the ancestors, who showed people how to care for country. Even as Aboriginal people learn to work with cattle, and form new and creative connections, in the process they participate in the wildness of colonial settlement. Rose traces such wildness to the spread of an Indo-European culture involving cattle from the Near East and India. She argues that wherever cattle have gone, so too has war: "I now suggest, provocatively, that the conquest of Australia did not begin in 1788. It began about 10,000 years ago when our ancestors domesticated cattle and began a long and intermittent career of cattle herding and raiding.... At about 6000 BP (before the present) the Proto-Indo-European group began to branch out linguistically and spatially.... The pastoralists of the steppes used both [horses and ox-drawn carts], and when they turned their attention to conquest they were devastatingly successful" (Rose 2004, 74–75).

The cattle-based war machine also has antistate possibilities. Consider the Nuer of South Sudan, as described in Evans-Pritchard's *The Nuer* (1940). The Nuer have a long and impressive history of pushing back the state, slave raiders, and even ordinary traders, who spin the threads of state and commercial power. According to Evans-Pritchard, they did so by exploiting the shifting ecology of their area, and particularly its inaccessibility during the rainy season, when much of the land is covered with water. This involves knowledge of wet and dry patches; Nuer alternated between dispersal and gathering as they led their herds to good grazing. They also used their political autonomy—and augmented it—by becoming expert cattle raiders, preying on their neighbors, the Dinka. During the period of Evans-Pritchard's study, much of Nuer politics involved the incorporation of Dinka war captives, grazing areas, and cows.

The mobility of cattle protected their autonomy, which allowed them to mount unified campaigns without authoritarian leaders. This is Deleuze and Guattari's war machine exemplified: the Nuer used their cattle-based mobilizations to repulse the state, with its demands for property and taxes, its settlement, and its surveillance. It is difficult not to read news of more recent mobilizations in South Sudan, such as the Nuer White Army, within this history of cattle-based antistate militarism (e.g., Young 2007). The Nuer have so far—incredibly—held off the state.

SEVEN. *Refusals of domestication are often problems of political ecology, not human ingenuity or a species' unwillingness to form relations with humans.*

What might a mushroom tell us about domestication? Matsutake are a cluster of *Tricholoma* species enjoyed as a gourmet food in Japan and Korea (Tsing 2015). Since the late 1970s, environmental changes have made them rare in Japan. Rarity raised prices; the Japanese economic boom made fine matsutake an exemplary gift. Thus, extraordinary efforts began to produce a domestic matsutake. Over the twenty years straddling the turn of this century, millions of yen were spent, high-power laboratories set up, and scientists gathered and deployed, all to discover the secrets of human control. Government grants as well as private companies supported the search. In the 1990s, the price of matsutake sometimes surpassed US$1000 per kilogram. At least until the price dropped with oversupply, a patented domestic matsutake would be worth millions.

Yet the research-and-development drive was a failure. Most of the sophisticated laboratories have now shut or are moving on to other research fields, perhaps with matsutake research as a sideline. A few advances were made, especially in technique, but no one succeeded in producing a matsutake mushroom in a laboratory.[10]

Ironically, matsutake in Japan is *already* intertwined with human affairs—and has been since the eighth century, when its delightful "autumn aroma" first entered a budding new Japanese literature. Matsutake lives with red pine, which in Japan grows mainly in the wake of human forest disturbance. The first mention of matsutake follows the deforestation of central Japan for temples and palaces and for fuel for iron forges. In the deforested countryside, Japanese red pines grew up as a weed, and with them matsutake. As state control spread across the Japanese islands, so did deforestation, pine, as a weed, and matsutake. Rice paddies replaced forests in valleys, but mountains were often left as peasant woodlands. Every time people cleared trees, pines came in as a weed—and with pine, matsutake. Japan's industrialization in the late nineteenth and early twentieth century produced plenty of pine, and plenty of matsutake. Logging for military needs was great for matsutake. The changes that made matsutake rare in the 1970s had more to do with the withdrawal of humans from the countryside than with their presence. Matsutake in Japan is a human commensal and cannot flourish without humans. Matsutake researchers have many ideas about how to make landscapes in which matsutake thrive. But these are not offered the same funding as laboratory cultivation, which could be translated to plantation cultivation.

The domestication initiatives across the turn of this century were not geared toward bringing humans and matsutake into relationship, since they already were. Instead, they aimed for the industrialization of matsutake; only this counted as domestication. If matsutake could be produced in a laboratory, then they could be grown in otherwise sterile plantations and made into standalone, interchangeable commodities, producing assets for investors. Domestication here means discipline, not interspecies engagement. That matsutake has refused to cooperate, at least so far, highlights the peculiarity of this definition, which we increasingly require to identify "domestication" at all.

All this speaks to the dearth of new domesticates. Childe was right in a way: states and armies keep us from being inventive. But this is not because we no longer dream. Rather, we recognize ingenuity as significant these days only when it takes the form of industrial capitalism. Interspecies relations make history within the opportunities and constraints of this political economy.

EIGHT. *Alienation and accumulation are threats to life on earth.*

Thinking with Japan's matsutake woodlands makes a larger point. One of the reasons matsutake have become rare in Japan is that in the mid-twentieth century peasant woodlands were converted to plantations of two timber trees, sugi and hinoki. After World War II, growing timber seemed a good way to save foreign exchange for oil, the scarcity of which had drawn Japan into war. Rural people were abandoning the countryside as urban economies expanded. Making forests industrially productive seemed like progress. Through state encouragement and regulation, former peasant woodlands were converted en masse into timber plantations. Matsutake do not grow with sugi and hinoki.

The timber plantations did not turn out as advocates had hoped. For the first decade after the war, labor was cheap and timber expensive; planners designed the plantations for these conditions. Farmers close-planted trees by hand on steep slopes. By the time the trees were ready for hand thinning and pruning, the price of labor had risen. Worse yet, the construction industry had pressured the government to lift the ban on importing wood. By then, too, trading companies had organized to pull mass quantities of resources from abroad. Cheap logs from Southeast Asia flooded the market; Japanese timber was no longer worth much. It was too expensive to thin and even too expensive to harvest the wood. Overcrowded trees invited pests and diseases. Except in a few established timber areas, silviculture was abandoned, leaving plantations in ruins.

Neither rural nor urban people like these crowded and unsalable ruins. Dense and biologically homogenous, they are not good places for hiking. In their packed homogeneity, they spread clouds of pollen, causing allergies. The growth trajectory of their conversion encouraged an overpopulation of deer, which became village pests. They fill the landscape, exemplifying the problems of plantations.

They also allow me to explain what I mean by "alienation" and "accumulation." Alienation is estrangement from community, society, and world. Marx used the term to refer to the estrangement of workers from their work and what they made. I extend the term to include other living beings: it refers to our removal from the familiar entanglements of our lifeworlds. Alienation has been a mainstay of the commodification of plants and animals. For sugi and hinoki, it meant growing each in an otherwise sterile monocrop. These trees ordinarily grow among mixed broadleafs; to become plantation commodities, they were separated from their biological companions, whether plants, animals, fungi, or bacteria. The whole point of the plantation is to leave out those companions; alienation is a basic principle of the plantation. It is also what made those plantations into ruins as soon as active management was discontinued. Because of alienation, ruined plantations do not offer the ecological services of forests.

Capitalist accumulation is the gathering of wealth for investment. It is the process by which investment generates more resources for investment, that is, the economic use of profit. In the story above, one might follow trading companies, which mobilized in postwar Japan to make money by organizing foreign trade. By putting together supply chains, they came to control the supply of resources in a growing Japan; at one point, they were among the most capital-rich companies in the world. The trading companies oversaw the deforestation of Southeast Asia for Japan's construction industry. It did not matter to them that the Japanese government had just engineered a huge supply of domestic timber. The logic of accumulation drove them not to worry about trees but rather to follow the money. In this search for profits, forest ruins were made in both Japan and Southeast Asia. Japanese matsutake was only one victim in a cataclysm of ecological destruction that stretched around the world.

The problem is much more general. The hegemonic forms of today's domestication complex—the plantation and the feedlot—make it difficult to even imagine multispecies life, much less enact it. For capital, ruins are not a problem so long as the money can be followed somewhere else. What will happen when we run out of "somewhere elses"?

NINE. *Unintentional cultivation and domestication-as-rewilding offer hopeful alternatives in imagining multispecies life with humans as a component.*

The term "unintentional cultivation" comes from my conversation with a matsutake scientist in Kyoto. He pointed out that farmers make matsutake landscapes by opening the forest, allowing pines to grow. This is a kind of cultivation, he noted, even if it is not what we think of as domestication. That discussion stimulated my thinking about the cospecies making of landscapes. Human and not human, we all shape our environments. Allowing humans into that larger category of world-making alters, too, our hopes for domestication. Other creatures are equally involved in "unintentional cultivation." What if we redefined the "homes" we wanted to make with other creatures to include the worlds they make as well? Might we come up with better ideas than industrial ruins?

"Domestication-as-rewilding" is an even more awkward term, pointing to how difficult it is to begin the productive confusion of domestic and wild—and our too easy naturalization of the domus. Social scientists always critique "the wild" for its ideological baggage; it's about time we extend the favor to "the domestic." The term I'm looking for, then, points to cospecies landscapes in which no species can be said to be in charge. Rather than alienation and accumulation, the point is multispecies engagement. One last matsutake story illustrates.

The Matsutake Crusaders are a group of volunteers who restore matsutake forests in abandoned peasant woodlands around Kyoto. Not all peasant woodlands were turned into timber plantations. Another problem for matsutake was the abandonment of woodlands by their human users. Before the 1950s, farmers used the woodlands for firewood and charcoal as well as a host of nontimber forest products, including matsutake. When fossil fuel use swept the nation, firewood and charcoal were dropped. Rural residents moved to the city, leaving only the elderly. The forms of woodland management that had maintained the earlier character of the landscape disappeared. New species grew up, changing the woodlands, which became dark and dense with evergreen broadleafs. Pines, which like light and mineral soils, were stressed and killed off by nematodes. Matsutake died with them.

In the late twentieth century, many Japanese came to worry about the changing character of the landscape. The term *satoyama*—the peasant woodland and its surrounding landscape of rice fields, irrigation channels, and gardens—gained currency. Citizen's movements organized to revitalize *satoyama*, which were seen not only as the source of Japan's beauty but also as a place where humans grew to be environmentally and socially aware. These groups organized to clear the new growth that followed the abandonment of the woodlands and to relearn coppicing and other arts for maintaining *satoyama*. They used hand tools and described their job as teaching and learning. Some volunteers were retired; some were students; others gave up weekends for this work.

The Matsutake Crusaders is one of many *satoyama* restoration groups. As with others, they advocate a principle of useful work. It is not enough to admire nature from afar, they say; people should immerse themselves in the *work* of producing good relations with their environments. Because of the high prices of matsutake, any mushrooms that grow on revitalized woodlands help support the process. Matsutake symbolize both the income-generating aspects of useful work and the pleasures of eating.

Matsutake Crusaders are aware that they may never see mushrooms on the hillsides they laboriously work. This is part of the appeal. Rather than forcing the mushrooms to appear, their goal is to make woodlands in which interspecies relations thrive. They see themselves as restoring abandoned ruins to multispecies life.

How might their example help us think about alternatives to hegemonic forms of domestication? Consider one not-so-obvious point: there are many, many species engaged with humans in unintentional cultivation and domestication-as-rewilding. If we want to show the productivity of our interspecies relations, we might look to these less-than-fully cultivated places. Bacteria, fungi, plants, and animals: a majority of the species left on earth are here because they have figured out a way to live with humans—a domestication of humans for their purposes, if you will. Where plantations and feedlots have not spoiled the land, water, and air for common use, we have plenty of partners in unintentional cultivation.

It is true that there are some extant species that do not stand human activities well. The spotted owl of western North America can nest only in old-growth forest; any tree cutting reduces the places it can live. The deep-sea tubeworms around volcanic vents have probably not reacted too much yet to our acidification of the oceans. I want to stand up for those species, too. But for the moment, let me turn to the species that manage to live in the messes humans make. Matsutake is one; so is pine. They are our autonomous codomesticates, working our landscapes, making worlds in which we both can live. If we value multispecies life on earth, we might start by appreciating their help.

This, as I'm arguing, is a domestication-as-rewilding that matters. Perhaps, in the spirit of Childe, we should plan a revolution around it. We could call it the Cospecies Accommodation Revolution (CAR). It's unlikely, given everything poised against it. Still, imagine this optimistic account of world history: "When capitalism left us in ruins, the CAR helped save a few patches of livability. Here were new forms of domestication, also known as rewilding, in which humans deferred to their multiple companion species as to desirable landscape modifications. In contrast to Noah's domestication on the ark, in the CAR many species built these not-very-homelike landscapes through their intersecting activities. Life on earth continues to exist because of these common efforts."

NOTES

1. I am grateful to Marianne Elisabeth Lien for soliciting this paper and for offering comments for its revisions. Heather Anne Swanson was also a generous interlocutor. Our dialogue as the volume developed may have made some points now seem repetitive; in hopes that they further the conversation, I have not removed them.

2. Many scholars have devoted careers to domestication; I am just a commentator, renarrating their work. For other novices, I recommend the lively and informative overviews offered in Smith's *The Emergence of Agriculture* (1995), Mithen's *After the Ice* (2004), and Scott's *Against the Grain* (2017).

3. Pigs, including the Bornean bearded pig and its domestic associates, interbreed across "species" lines rather readily. Recent research suggests that pig domestication occurred in many independent sites, involving varied wild pigs; in each area "domestic pigs" have genes from nearby wild stock (Larson et al. 2005).

4. The nutritional reduction and zoonotic diseases of domestic encampments reduced everyone's stature, male and female.

5. This once common claim is beginning to disappear from popular sources. But see, for example, Mescher (2005).

6. See Mintz (1974) for Puerto Rico.

7. Elaine Gan (2016; personal communication) notes that most of the varietal diversity bred into commercial crops has a very narrow genetic range inherited from a few chosen ancestors.

8. As Alves (2011, 73) puts it, "To be a Spanish conqueror was to be on the move with other animals."

9. See, for example, http://www.theguardian.com/world/2009/apr/08/brazilian-murder-dorothy-stang.

10. Matsutake are ectomycorrhizal fungi, living through joining with the roots of plants, from which they get their carbohydrates. Researchers had two choices for producing matsutake in the laboratory: they might find a strain that was willing to change this livelihood preference to feed from dead organic matter or they might find a way to bring mycorrhizal connections into the laboratory, working to control the plant-fungus interface. Matsutake researchers worked on both. Neither strategy offered easy results.

REFERENCES

Alves, Abel. 2011. *The Animals of Spain: An Introduction to Imperial Perceptions and Human Interaction with Other Animals, 1492–1896.* Leiden, Netherlands: Brill.

Anderson, Virginia de John. 2006. *Creatures of Empire: How Domestic Animals Transformed Early America.* New York: Oxford University Press.

Cassidy, Rebecca, and Molly Mullin, eds. 2007. *Where the Wild Things Are Now: Domestication Reconsidered.* Oxford: Berg.

Childe, V. Gordon, 1951 [1936]. *Man Makes Himself.* New York: New American Library.

Day, Anthony, 1994. "'Landscape' in Early Java." In *Recovering the Orient: Artists, Scholars, Appropriations*, ed. A. Gerstle and A.C. Milner, 175–203. Chur, Switzerland: Harwood Academic Publishers.

Deleuze, Gilles, and Félix Guattari. 1993. *A Thousand Plateaus.* Minneapolis: University of Minnesota Press.

Elson, Robert. 1994. *Village Java under the Cultivation System, 1830–1870.* Sydney: Allen and Unwin.

Evans-Pritchard, E. E. 1940. *The Nuer.* Oxford: Clarendon.

Ficek, Rosa. 2014. "The Pan American Highway: An Ethnography of Latin American Integration." PhD dissertation, University of California, Santa Cruz.

Fuller, Dorian Q. 2007. "Contrasting Patterns in Crop Domestication and Domestication Rates: Recent Archaeobotanical Insights from the Old World." *Annals of Botany* 100: 903–24.

Gan, Elaine. 2016. "Time Machines: Coordinating Change and Emergence." PhD dissertation, University of California, Santa Cruz.

Harlan, Jack, 1992. *Crops and Man.* 2nd ed. Madison, WI: American Society of Agronomy-Crop Science Society.

Hodder, Ian, 1990. *The Domestication of Europe.* Oxford: Basil Blackwell.

Larson, Greger, Keith Dobney, Umberto Albarella, Meiying Fang, Elizabeth Matisoo-Smith, Judith Robins, Stewart Lowden, Heather Finlayson, Tina Brand, Eske Willerslev, Peter Rowley-Conwy, Leif Andersson, and Alan Cooper. 2005. "Worldwide Phylogeography of Wild Boar Reveals Multiple Centers of Pig Domestication," *Science* 307, no. 5715: 1618–21.

Leach, Helen, 2007. "Selection and the Unforeseen Consequences of Domestication." In *Where the Wild Things Are Now*, ed. Cassidy and Mullin, 71–100.

Mescher, Virginia. 2005. "How Sweet It Is! A History of Sugar and Sugar Refining in the United States." http://www.raggedsoldier.com/sugar_history.pdf.

Mintz, Sidney. 1974. *Worker in the Cane.* New York: W. W. Norton.

Mithen, Steven. 2004. *After the Ice: A Global Human History, 20,000–5,000 BC.* Cambridge, MA: Harvard University Press.

Porcher, Richard, and Sarah Fick. 2005. *The Story of Sea Island Cotton.* Layton, UT: Gibbs Smith.

Rose, Deborah Bird. 2004. *Reports from a Wild Country.* Sydney: University of New South Wales Press.

Scott, James. 2009. *The Art of Not Being Governed.* New Haven: Yale University Press.

Scott, James. 2017. *Against the Grain: A Deep History of the First Agrarian States.* New Haven: Yale University Press.
Smith, Bruce. 1995. *The Emergence of Agriculture.* New York: Scientific American Library.
Tsing, Anna. 1995. *In the Realm of the Diamond Queen: Marginality in an Out-of-the-Way Place.* Princeton: Princeton University Press.
Tsing, Anna. 2003. "Cultivating the Wild." In *Culture and the Question of Rights*, ed. Charles Zerner, 24–55. Durham: Duke University Press.
Tsing, Anna. 2012. "On Nonscalability: The Living World Is not Amenable to Precision Nested Scales." *Common Knowledge* 18, no. 3: 505–24.
Tsing, Anna. 2015. *The Mushroom at the End of the World: On the Possibility of Life in Capitalist Ruins.* Princeton: Princeton University Press.
Ucko, Peter, and G. W. Dimbleby. 1969. "Introduction." In *The Domestication and Exploitation of Plants and Animals.* Chicago: Aldine. Proceedings of the research seminar in archaeology and related subjects, 1968, xvii–xxii.
Wells, Jonathan, Jeremy DeSilva, and Jay Stock. 2012. "The Obstetric Dilemma: An Ancient Game of Russian Roulette, or a Variable Dilemma Sensitive to Ecology?" *Yearbook of Physical Anthropology* 149: 40–71.
Wilson, Peter, 2007. "Agriculture or Architecture? The Beginnings of Domestication." In *Where the Wild Things Are Now*, ed. Cassidy and Mullin, 101–22.
Young, John. 2007. "The White Army: An Introduction and Overview." A Working Paper of the Sudan Human Security Baseline Assessment Project of the Small Arms Survey, http://www.smallarmssurveysudan.org/fileadmin/docs/working-papers/HSBA-WP-05-White-Army.pdf.

Contributors

INGER ANNEBERG is a postdoctoral researcher in the Department of Animal Science at Aarhus University, Denmark, and has a background as a journalist in addition to a PhD in Communication (2012), during which she focused on issues of animal welfare. She studies the actions of and interactions between authorities and livestock farmers in relation to animal welfare legislation, and her research focus is how the concept of animal welfare is handled and interpreted among different groups, for instance, farm employees, students at agricultural colleges, farm owners, and state authorities.

NATASHA FIJN is a Fejos Postdoctoral Fellow in Ethnographic Film, funded by the Wenner-Gren Foundation for the purpose of filming the medical treatment of herders and herd animals in Mongolia in 2017. Her research engages aspects of visual anthropology, animal domestication, and animal studies. Her ongoing interest is in cross-cultural perceptions and attitudes toward other animals and in using the visual, particularly observational filmmaking, in her research. Her book *Living with Herds: Human-Animal Coexistence in Mongolia* was published by Cambridge University Press in 2011.

RUNE FLIKKE is an associate professor in the Department of Social Anthropology at the University of Oslo. He has extensive research experience with African Independent Churches in Durban, South Africa, and has participated in a four-year research project on vaccination in Malawi. In addition to a growing engagement with resource management and nature conservation in general, he is currently researching the significance of conceptions of air and atmosphere in the context of landscape alterations in South Africa.

FRIDA HASTRUP is an associate professor of ethnology at the Saxo Institute, University of Copenhagen. From 2013 to 2016, she was the leader of a collaborative research project about natural resources that explored ethnographically how these get made and unmade. In 2015–16, she was a member of the Arctic Domestication in the Era of the Anthropocene research group at the Center for Advanced Study in Oslo, Norway. She has published widely in the field of environmental anthropology, including the monograph

Weathering the World: Recovery in the Wake of the Tsunami in a Tamil Fishing Village (2011).

MARIANNE ELISABETH LIEN is a professor in the Department of Social Anthropology at the University of Oslo and was the director of the research group Arctic Domestication in the Era of the Anthropocene at the Center for Advanced Study in Oslo (2015–16). She is concerned with relations that connect humans, animals, and their environment, with a focus on the Scandinavian Arctic and Tasmania. Lien has published widely on food, consumption and marketing, nature engagement, invasive species, salmon aquaculture, animal welfare, and domestication. Her latest book is *Becoming Salmon: Aquaculture and the Domestication of a Fish* (2015).

KNUT G. NUSTAD is a professor and the head of the Department of Social Anthopology at the University of Oslo, where he works with political and environmental theory broadly defined. He has conducted research on informal political processes in urban settlements, development policy, and state formation as well as land reform, conservation and protected areas, with South Africa as his primary empirical focus. His latest book is *Creating Africas: Struggles over Nature, Conservation and Land* (2015).

JON HENRIK ZIEGLER REMME is an associate professor in social anthropology at the University of Oslo. His research is concerned with relations between humans, spirits, plants, and animals among the Ifugao of Northern Luzon, the Philippines. Remme is currently exploring the politics, ethics, and economy of multispecies relations in the lobster industry in Norway and Maine, USA. Remme has published on a variety of themes, including Ifugao sacrificial rituals, agricultural practices, and Pentecostalism. His books include *Pigs and Persons in the Philippines: Human-Animal Entanglements in Ifugao Rituals* (2014) and the edited volume *Human Nature and Social Life: Perspectives on Extended Socialities* (2017).

SARA ASU SCHROER is a research fellow in the interdisciplinary ERC project Arctic Domus in the Department of Anthropology at the University of Aberdeen. Her current research deals with the relationships that develop between birds of prey and humans in the practice of falconry as well as in domestic breeding projects. Here, she is interested in the multispecies landscapes, technologies, and intimacies of avian domestication, using approaches from environmental anthropology, multispecies ethnography, and the anthropology of learning and enskilment.

HEATHER ANNE SWANSON is an associate professor of anthropology at Aarhus University, Denmark, where she is also the deputy director of the Centre for Environmental Humanities and a participant in the Aarhus University Research on the Anthropocene project. In 2015–16, she was also a member of the Arctic Domestication in the Era of the Anthropocene research group at the Center for Advanced Study in Oslo, Norway. In addition to her research on salmon and fisheries, Swanson has authored publications about diverse environmental and political concerns, including agricultural practices, railroad construction, and academic approaches to the Anthropocene. She is a coeditor of *Arts of Living on a Damaged Planet* (2017).

ANNA LOWENHAUPT TSING is a professor of anthropology at the University of California, Santa Cruz, and a Niels Bohr Professor at Aarhus University, where she codirects Aarhus University Research on the Anthropocene (AURA). She is the author of *The Mushroom at the End of the World: On the Possibility of Life in Capitalist Ruins* (2015) and the coeditor of *Arts of Living on a Damaged Planet* (2017). One of her emerging projects is as co-organizer of *The Feral Atlas*, an interactive digital media production concerning life out of whack, from invasive species to newly virulent pathogens.

METTE VAARST is senior researcher at Aarhus University, Department of Animal Science, with a background in veterinary science, focusing on animals, their farmers in their farm and community contexts, human choices, perceptions, communication and dialogues around animals, interactions and actions regarding farming, advisory service, health concepts and disease handling, the different roles of animals, and their connections and connectedness with humans. She had been educated as a classical homoeopath (1994) and veterinary homoeopath (1997), and after her PhD (1995), she earned an additional masters in health anthropology (2007) with studies in medical anthropology, ethnic medicine, and philosophy of science.

GRO B. WEEN is an associate professor in social anthropology at the Cultural History Museum, University of Oslo. At the museum, she is the head of the Department of Ethnography, Numismatics, Classic Archaeology, and University History, and keeper of the Arctic and Australian collections. Ween was a member of the research team of Lien's Arctic Domestication in the Era of the Anthropocene at the Oslo Academy of Sciences and Letters' Centre of Advanced Study. She has published extensively on nature practices, natural resource management, issues of cultural heritage, and indigenous politics.

Index

Page numbers in italics refer to photographs.